MEASURING HEALTH EQUITY IN SMALL AREAS – FINDINGS FROM DEMOGRAPHIC SURVEILLANCE SYSTEMS

Measuring Health Equity in Small Areas – Findings from Demographic Surveillance Systems

INDEPTH NETWORK

Taylor & Francis Group

LONDON AND NEW YORK

First published 2005 by Ashgate Publishing

Reissued 2018 by Routledge
2 Park Square, Milton Park, Abingdon, Oxon, OX14 4RN
52 Vanderbilt Avenue, New York, NY 10017

First issued in paperback 2020

Routledge is an imprint of the Taylor & Francis Group, an informa business

Publisher's Note
The publisher has gone to great lengths to ensure the quality of this reprint but points out that some imperfections in the original copies may be apparent.

Disclaimer
The publisher has made every effort to trace copyright holders and welcomes correspondence from those they have been unable to contact.

A Library of Congress record exists under LC control number: 2004028596

Typeset in Times Roman by N²productions

ISBN 13: 978-0-367-66726-9 (pbk)
ISBN 13: 978-0-8153-9048-0 (hbk)

Contents

Foreword

It is a great pleasure and honour for me to be asked to write a foreword to *Measuring Health Equity in Small Areas: Findings from Demographic Surveillance Systems*. The editors, along with the authors of the chapters and the numerous other contributors to this volume deserve tremendous credit for illuminating with the highest academic rigour the issues surrounding the complex terrain of inequalities and inequities in health and health care in some of the world's poorest areas.

The selection of INDEPTH sites as the common denominator for the various analyses is in itself an important statement related to equity. In the early years of the 21st Millennium, it is indeed worrisome that in many parts of the world we remain 'in the dark' about life's vital events – namely birth and death. In Sub-Saharan Africa, registration of these vital events remains incomplete in every country with only partial coverage in a small minority of countries. From this extreme, a strong association can be drawn between coverage of vital registration systems and the level of population health: those countries with no, or limited, registration have the poorest levels of health. This dissonance between the absence of health information and the need given the magnitude of health challenges has been referred to as the 'information paradox'. INDEPTH sites, through the demographic surveillance systems – accurate and continuous coverage of a total population's health – are an important step in redressing what I refer to as the 'first injustice' i.e. the lack of a right for one's life to be counted. Although significant as single sites, together the INDEPTH sites can pool their vital data to provide critical insights in the patterns of health in some of the world's poorest areas where health problems are the greatest and unfortunately the least understood or appreciated. More simply put, INDEPTH as an entity makes the critical statement that all lives count and deserve to be counted.

From this 'first right' – i.e. the right to be counted, this volume examines inequalities and inequities in health and access to health care within the various INDEPTH sites.

This activity constitutes an important challenge to health development orthodoxy, namely the assumption of homogeneity among the poor i.e. that all persons living in poor countries are equally poor and equally unhealthy. As the chapters in this volume make abundantly clear, there are very significant socio-economic, gender, racial-ethnic, place of residence, occupation and religious stratifiers of health status and access to services in areas where heretofore there was little or no awareness. The data from the Rufiji DSS in Tanzania (Mwageni et al.) of significant child mortality and access to bednet inequalities are indicative of the dangers of assuming that all rural poor are equally unhealthy. The nature of the inequality both in its magnitude e.g. steepness of gradients, and the patterns according to specific stratifiers and outcome measures is variable (and in some cases not significant) and therefore defies simplistic generalizations. The patterns of distribution also raise important questions related to measurement e.g. the consumer expenditure proxy (CEP) in Tanzania (Setel et al.) that in some cases require further verification or supplemental research that moves more to causes. The finding that children of fathers with the highest social support had increased mortality in a setting where socioeconomic status appears not to be related to child mortality raises a number of interesting questions relating to social support and the health care of children (Ratcliffe et al.). Beyond the inherent value of this exercise in terms of developing critical competencies and capacities for distributional analysis, the praxis of understanding variation holds much promise in terms of developing more appropriate policies and programs and understanding what does (and doesn't) work.

The volume's ambition doesn't stop, however, at the mere description of distribution, or inequalities, rather it goes beyond and asks the more complex questions about 'fairness of distribution'. Although, in general, the language of distribution is rather confused on the distinction between judgements about fairness and measures of empirical reality, the use of the word 'equity' is unambiguous in terms of invoking a moral judgement. The INDEPTH volume makes some inroads in moving towards the measurement of health inequities raising some inherent challenges. Foremost among these is the standard, or fair, distribution against which one is assessing unfairness. Classically, it is assumed that fairness in distribution is one of equality: groups have equal health outcomes or equal access to care. However, recognition of different norms or standards for health suggests that such assumptions may be incorrect. For 'gender', for example, many would argue that we should expect women to have a longer life expectancy than men representing a biological or inherent difference between the sexes and hence equality in life expectancy between the sexes may be inequitable. In another case, evidence that HIV positive persons in all socio-economic groups have equally 'poor' access to anti-retrovirals may not be evidence of equity but rather significant inequity if there is an expectation that all groups should have 'good' access to anti-retrovirals. Hence the issue of the fairness benchmark against which to measure inequity is but one of the critical challenges raised in this volume.

Beyond theory and research, however, the practical and applied nature of the analyses in this book ensures that this volume will not only sit well on the shelves but 'walk the talk' on the ground. Across a diverse group of countries and settings, seminal insights and opportunities for action abound:

- The poorest mothers in Senegal rely much more on underage caretakers for the sick child and do significantly less to prevent illness (Sodemann et al.)
- Aggressive social marketing of bed nets in a rural district of Tanzania has diminished the poorest/least poor ratio in ownership (Nathan et al.)
- Impressive declines in excess mortality among girls aged 1-4 years over a 10 year period in rural Bangladesh (Razzaque et al.)
- Education of mothers, and not wealth of the households, is more strongly associated with higher child survival in rural Ghana (Debpuur et al.).

From case studies and solid analyses like those in this volume, the challenges of health and health care become grounded in the pervasive reality of unequal social risks and opportunities. Health researchers and practitioners must not only be sensitive to inequalities but learn to understand and redress them more effectively to enhance prospects for more equitable health development and health systems performance. The INDEPTH network in making the values of fair distribution more explicit, and in measuring and analysing inequalities is taking critical steps to insuring that health progress is evidence-based and sensitive to the needs of those with the greatest needs.

Dr Timothy Evans
Assistant Director-General
Evidence and Information for Policy
World Health Organization

Preface

This publication results from collective exploratory research efforts in the area of health equity funded by the INDEPTH Network with support from the Rockefeller Foundation. Ten member sites in sub-Saharan Africa and Southeast Asia participated in a coordinated program of research to explore, adapt and apply appropriate demographic surveillance system (DSS) survey methods to determine the relationship between socio-economic factors at multiple levels and inequities of both access to health services, and health outcomes, in order to assist programme and policy makers to overcome health status disparities and improve overall health status of the poor. Specifically, the studies examined how assets, consumption expenditure, gender, education, occupation, social connectivity and other socio-economic status proxies (e.g. housing and water source) relate to mortality and service use in various population subgroups.

Over the past decade, several initiatives have been launched to address the major health problems affecting the world's poorest countries, including global efforts to combat HIV/AIDS, TB and malaria. More recently, a millennium challenge has been laid down to root out and confront the links between poverty and health. While there has been a spate of studies and reports devoted to this subject, it has been noted that policies meant to address the needs of the global poor are often based on indirect estimates and data from urban centres and health facilities that may not accurately reflect the experience of the poorest. In 2000, Davidson Gwatkin attended an INDEPTH Annual Scientific Conference and pointed out that new methods of socio-economic stratification that are applicable to household surveys had recently been developed. He strongly encouraged INDEPTH member sites to look into this. Since INDEPTH DSS sites collectively represent one of the largest, continuous, longitudinal sources of household and health outcomes data in the poorest countries, it made sense to examine these developments and determine to what extent they might be integrated into DSS. Using DSS, the INDEPTH researchers aim to contribute both to the empirical knowledge about health equity in developing countries and to report on the application of and innovation in tools and methods. More importantly, they want to strengthen their ability to monitor progress to redressing both the intolerable burden of disease among the poor, and the inequities they suffer.

Illustrated with case studies from sub-Saharan Africa and Asia, this book puts forward a comprehensive view of the methodologies and findings of INDEPTH member sites to date. It develops and measures concepts and constructs of 'poverty' and 'equity' and relates these to health status and health system access. This volume also grapples with new concepts and tools to measure changes in deprivation and disadvantage, adding to this intense theoretical and methodological debate.

The contributions of the INDEPTH Health Equity project are unique in the following ways: they have measured health inequities in small geographic areas, which has too seldom previously been done, and; they have demonstrated the feasibility of undertaking health equity assessments in typical, ongoing demographic and epidemiological surveillance on a vast array of topics (as distinct from field studies designed specifically for health equity measurement from the outset). The consistent findings that clear equity gradients exist even within small rural populations is another important contribution of the research presented here. The Network was fortunate to have the support of several international experts for this first cross-site attempt to measure health equity in demographic surveillance areas. These experts have contributed introductory and concluding chapters (Chapters 1 and 12) that add international perspectives beyond the contributions of the INDEPTH member sites. Both chapters provide a synthesis of the work carried out by the sites. The introduction assesses the methodologies employed by the researchers to measure socio-economic status of the populations studied in the small areas; the conclusion deals with the collective significance of the INDEPTH results.

Given the existence of health inequities as demonstrated in the site chapters, there is little surprise that INDEPTH has now moved its health equity work to a second phase. It has developed a standard poverty measurement tool that may be used in DSS sites or other data collection platforms (http://www.indepth-network.net/ core_documents/indepthtools.htm). Recently, the Network has funded nine of its member sites to develop equity intervention studies to assess whether selected existing health interventions or services can be targeted to achieve a greater pro-poor focus. It is hoped that current Network efforts will lead to a sequel of this publication that goes beyond merely measuring inequity to actually reducing it.

Editors
Don de Savigny, (Coordinating Editor; Swiss Tropical Institute, Basel; Formerly Rufiji DSS and Tanzania Essential Health Interventions Project)
Cornelius Debpuur, (Navrongo DSS, Ghana)
Eleuther Mwageni, (Rufiji DSS, Tanzania)
Rose Nathan, (Ifakara DSS, Tanzania)
Abdur Razzaque, (Matlab DSS, Bangladesh)
Philip W Setel, (MEASURE Evaluation, Carolina Population Center, University of North Carolina at Chapel Hill; Formerly AMMP DSS, Tanzania)

March 2005

Acknowledgements

Thanks to initial funding from the Rockefeller Foundation (USA) and subsequent contributions from the World Bank (USA), Sida/SAREC (Sweden) and the Wellcome Trust (UK), the INDEPTH Network established in 2002 its first study in the area of health equity. Project reports from site-specific studies have formed the basis for this publication. The INDEPTH Network would therefore like to acknowledge the authors of site chapters in this publication as well as the various staff at the sites who contributed to the successful completion of the studies.

The INDEPTH Network further acknowledges the individual and various contributions of the editorial team for managing the publication, several referees who provided expert guidance to sites, site leaders and their teams for their untiring efforts in addressing reviewers' comments, and the INDEPTH Secretariat for its efficient co-ordination.

We wish to express special thanks to Davidson Gwatkin (World Bank) and Saul Morris (DFID, formerly London School of Hygiene and Tropical Medicine) for contributing chapters to this book and for facilitating at workshops on writing the reports.

The Network would like to thank Fred Binka for overall co-ordination of this work, and staff in the INDEPTH Secretariat (Felicia Manu, Sixtus Apaliyah, Titus Tei, Kwabena Owusu-Boateng, Ayaga Bawah, and Osman Sankoh) for diverse support provided.

Finally, INDEPTH would like to acknowledge the efficiency and supportive manner in which the Ashgate Publishing team in the UK handled the publication process of this book.

INDEPTH Network

List of Figures

List of Tables

List of Annexes

List of Contributors

Chapter 1

Saul S. Morris   London School of Hygiene and Tropical Medicine, London, UK

Chapter 2

Eleuther Mwageni   Rufiji Demographic Surveillance System, Tanzania; Sokoine University of Agriculture, Tanzania

Honorati Masanja   Tanzania Essential Health Interventions Project (TEHIP), Tanzania; Ifakara Health Research and Development Centre (IHRDC), Tanzania

Zaharani Juma   Rufiji Demographic Surveillance System, Tanzania

Devota Momburi   Rufiji Demographic Surveillance System, Tanzania

Yahya Mkilindi   Rufiji Demographic Surveillance System, Tanzania

Conrad Mbuya   Tanzania Essential Health Interventions Project (TEHIP), Tanzania; Ministry of Health, Tanzania

Harun Kasale   Tanzania Essential Health Interventions Project (TEHIP), Tanzania; Ministry of Health, Tanzania

Graham Reid   Tanzania Essential Health Interventions Project (TEHIP), Tanzania; International Development Research Centre (IDRC) Canada

Don de Savigny   Tanzania Essential Health Interventions Project (TEHIP), Tanzania; International Development Research Centre (IDRC) Canada

Chapter 3

Rose Nathan   Ifakara Demographic Surveillance System; Ifakara Health Research and Development Centre, Ifakara, Tanzania

Joanna Armstrong-Schellenberg   Ifakara Health Research and Development Centre, Ifakara, Tanzania; Gates Malaria Partnership, London School of Hygiene and Tropical Medicine, London, UK

Honorati Masanja   Tanzania Essential Health Interventions Project (TEHIP), Tanzania; Ifakara Health Research and Development Centre (IHRDC), Tanzania

Sosthenes Charles   Ifakara Health Research and Development Centre, Ifakara, Tanzania

Oscar Mukasa   Ifakara Health Research and Development Centre, Ifakara, Tanzania

Hassan Mshinda	Ifakara Health Research and Development Centre, Ifakara, Tanzania

Chapter 4

Cornelius Debpuur	Navrongo Demographic Surveillance System; Navrongo Health Research Centre, Navrongo, Ghana
Peter Wontuo	Navrongo Health Research Centre, Navrongo, Ghana
James Akazili	Navrongo Health Research Centre, Navrongo, Ghana
Philomena Nyarko	University of Ghana, Legon, Ghana

Chapter 5

Kathleen Kahn	Agincourt Demographic Surveillance System; School of Public Health, University of the Witwatersrand, Johannesburg, South Africa
Mark Collinson	School of Public Health, University of the Witwatersrand, Johannesburg, South Africa
James Hargreaves	School of Public Health, University of the Witwatersrand, Johannesburg, South Africa
Sam Clark	School of Public Health, University of the Witwatersrand, Johannesburg, South Africa
Stephen Tollman	School of Public Health, University of the Witwatersrand, Johannesburg, South Africa

Chapter 6

Morten Sodemann	The Bandim Demographic Surveillance System, Bandim, Guinea-Bissau
Amabelia Rodrigues	The Bandim Demographic Surveillance System, Bandim, Guinea-Bissau
Jens Nielsen	The Bandim Demographic Surveillance System, Bandim, Guinea-Bissau
Peter Aaby	The Bandim Demographic Surveillance System, Bandim, Guinea-Bissau

Chapter 7

Amy Ratcliffe	Farafenni Demographic Surveillance System, Medical Research Council, Farafenni, The Gambia
Kate Halton	Farafenni Demographic Surveillance System, Medical Research Council, Farafenni, The Gambia
Rosalind Coleman	Farafenni Demographic Surveillance System, Medical Research Council, Farafenni, The Gambia
Maimuna Sowe	Farafenni Demographic Surveillance System, Medical Research Council, Farafenni, The Gambia
Gijs Walraven	Farafenni Demographic Surveillance System, Medical Research Council, Farafenni, The Gambia

Chapter 8

Nguyen Duy Khe	FilaBavi Demographic Surveillance System; MCH/FP Department, Ministry of Health, Hanoi, Vietnam; IHCAR, Department of Public Health Sciences, Karolinska Institutet, Stockholm, Sweden
Pham Huy Dung	Health Strategy and Policy Institute, Hanoi, Vietnam
Ho Dang Phuc	Institute of Mathematics, Hanoi, Vietnam
Hoang Van Minh	Hanoi Medical University, Hanoi, Vietnam
Nguyen Xuan Thanh	Hanoi Medical University, Hanoi, Vietnam
Bo Eriksson	Nordic School of Public Health, Gothenburg, Sweden
Vinod Diwan,	IHCAR, Department of Public Health Sciences, Karolinska Institutet, Stockholm, Sweden; Nordic School of Public Health, Gothenburg, Sweden
Nguyen Thi Kim Chuc	IHCAR, Department of Public Health Sciences, Karolinska Institutet, Stockholm, Sweden; Hanoi Medical University, Hanoi, Vietnam

Chapter 9

Abdur Razzaque	Health and Demographic Surveillance System ICDDR, B, Centre for Health and Population Research, Dhaka, Bangladesh
Peter Kim Streatfield	Health and Demographic Surveillance System ICDDR, B, Centre for Health and Population Research, Dhaka, Bangladesh

Chapter 10

Abdullahel Hadi	Watch Project Demographic Surveillance System, Research and Evaluation Division, BRAC, Dhaka, Bangladesh
M.Showkat Gani	Watch Project, Research and Evaluation Division, BRAC, Dhaka, Bangladesh

Chapter 11

Philip Setel	Dar es Salaam, Hai, and Morogoro Demographic Surveillance Systems; University of Newcastle-upon-Tyne; Adult Morbidity and Mortality Project
Savitri Abeyasekera	Statistical Services Centre, University of Reading
Patrick Ward	Oxford Policy Management
Yusuf Hemed	Adult Morbidity and Mortality Project
David Whiting	University of Newcastle-upon-Tyne; Adult Morbidity and Mortality Project
Robert Mswia	Adult Morbidity and Mortality Project
Manos Antoninis	Oxford Policy Management

Chapter 12

Davidson R. Gwatkin	The World Bank, The Rockefeller Foundation

Chapter 1

Epidemiology and the Study of Socio-economic Inequalities in Health

Saul S. Morris

Background

The study of socio-economic inequalities in health offers unique and exciting opportunities for collaboration between epidemiologists and social scientists. These collaborations will take place against a background of increasing numbers of articles with socio-economic status keywords in their titles being published in health journals; such articles already numbered over 200 per year in the late 1990s (Oakes and Rossi, 2003). Interest in socio-economic status can be seen as one aspect of a broader 'critical reengagement' between epidemiology and the social sciences (Krieger, 2000) and the development of a new sub-discipline of 'social epidemiology', heralded in numerous editorials (Krieger, 1999; PAHO, 2002; Kasl and Jones, 2002; Kawachi, 2002).

Socio-economic status has not always enjoyed such an exalted position in epidemiologic research. Most students of the discipline will have been introduced to the concept as a 'confounder', a variable that distorts the analysis of causal relationships because of its association with both disease outcomes and exposure to 'true' bio-medical risk factors. Economists, on the other hand, might see income and ownership of assets as more malleable to policy intervention than pills, immunizations, or other procedures that require elaborate health systems to deliver. Perhaps the shared experience of the 1990s – which saw profound economic transformations in many countries followed by equally striking epidemiologic changes – may temper these polarized views and facilitate a consensus around the web of causality in which bio-medical and socio-economic variables interact.

In spite of the heightened interest, there remains a pervasive feeling in the health community that socio-economic status is a difficult-to-pin-down concept requiring elaborate and unfamiliar measurement techniques. This leads some epidemiologists to feel uncomfortable about attempting to quantify socio-economic differentials in health outcomes and understand how they are generated. In this chapter, we examine how the different INDEPTH demographic surveillance sites have managed to develop relatively simple measures of socio-economic status, based on limited datasets and – for the most part – eschewing highly sophisticated analytic techniques. We start by asking why more familiar constructs such as income are not generally used in health surveys in developing countries, and then proceed to question what is actually meant by 'socio-economic status'. We then review how socio-economic status has been conceptualized and measured in the different studies presented in this volume, together with some of the alternative methods that are popular elsewhere in the literature. Finally, we examine what is known about the technical properties of alternative measures, and whether concerns about these technical properties are sufficient to affect policy conclusions about the importance of socio-economic differentials in health outcomes. Throughout, the discussion in this chapter is limited to quantitative methods that can be applied across large populations; participatory techniques that require personal knowledge of all households being classified are not considered.

The Problem with Income

In principle, income would seem to be a good measure of socio-economic status because it bears a straightforward relation to the minimum level of resources needed to participate fully in society. It is widely used both in census and in survey work in developed countries, and often forms an important part of 'integrated' household surveys conducted in developing countries – such as the World Bank's series of Living Standard Measurement Surveys (www.worldbank.org/lsms/). For routine use in research on health inequalities, however, income suffers from a number of conceptual and practical disadvantages that have previously been described in some detail by Falkingham and Namazie (2002).

Firstly, the very idea of reducing the complex concept of poverty to a money metric is disputed, as this relegates non-monetary aspects of poverty such as social exclusion to a secondary status. Secondly, difficulties arise in dealing with leisure time: if an individual chooses to enjoy leisure rather than work, this should presumably not be considered to indicate a lower level of welfare than would apply if the same individual reversed that decision and worked all the hours available to them. Thirdly, the same amount of income is clearly not as adequate for a family of ten as it is for a single individual, but the common practice of dividing household income by the number of household members (to obtain *per capita* income) also seems misleading because it ignores 'economies of scale'.[1] Fourthly, income has been observed to fluctuate over the course of a year, far more than the living standards it is supposed to represent (because individuals are able to save and draw down savings to compensate for short-term fluctuations in income).

Even in the absence of these conceptual difficulties, actually measuring income in developing countries is extremely difficult. This is because huge numbers of people in poor countries are self-employed, and their 'income' has to be painstakingly reconstructed by subtracting their business costs from the value of their production. In many cases, their production is never actually sold, so not even this quantity is known, requiring multiple and rather complex imputations to arrive at the true value. In effect, it is necessary to impose an accounting framework on all the household's transactions (as described by Johnson, McKay and Round, 1990), and ask a large number of

---

1   The costs of maintaining constant levels of individual welfare are not likely to increase in direct proportion to family size, both because some household level goods can be used by all household members at no extra cost, and because large families tend to have more children, and children do not consume as much as adults.

probing questions to obtain the information needed. Applying such a questionnaire in the field can take an hour or more, and the questions asked are often perceived to be invasive.

In the light of these difficulties, it has been suggested that 'in the context of measuring welfare in developing countries, there is a very strong case in favor of using measures based on consumption not income' (Deaton, 1997). In this case, 'consumption' means the full value of all goods and services consumed by an individual or household, which is thought to fluctuate much less than income. Nonetheless, there are still many conceptual difficulties (once again described in detail by Falkingham and Namazie, 2002), and actually collecting information on consumption remains a hugely demanding task because of the need to disaggregate all the different goods and services consumed to facilitate recall. Furthermore, Sahn and Stifel (2003) have pointed out that 'there is considerable merit in moving the process of poverty measurement away from solely expenditure-based measures' because it is the distribution of productive assets that needs to be addressed to achieve 'meaningful poverty alleviation'. In the following section we therefore consider measures of socio-economic status that take us beyond an exclusive focus on income and consumption.

The Meaning of 'Socio-economic Status'

While related concepts such as 'wealth' or 'social class' have been rather rigorously studied, and enjoy some consensus with respect to their meanings, it is less clear exactly what is intended by the term 'socio-economic status'. Indeed, Krieger, Williams and Moss (1997) have gone so far as to reject this term entirely, on the grounds that it 'blurs distinctions between two different aspects of socio-economic position: (a) actual resources, and (b) status, meaning prestige- or rank-related characteristics'. Even if this is the case, it seems unlikely that the term will disappear in the near future, and a simple, unambiguous definition would therefore be helpful. One such definition is proposed by Oakes and Rossi (2003), who view socio-economic status as 'differential access (realized and potential) to desired resources'. These resources are understood to fall into three distinct domains:

> (1) material endowments (e.g. earned income, investment income, real property, and other fungible goods), (2) skills, abilities and knowledge, and (3) one's social network and the status, power, trustworthiness, and abilities of its members (Oakes and Rossi, 2003, p. 776).

These three domains are given the labels of 'material capital', 'human capital', and 'social capital'. The task of measuring socio-economic status is thus 'reduced' to identifying appropriate indicators for each of these domains, and summarizing the information in a single metric defined at an appropriate level – that of the individual, household, or community. Few authors have attempted to be more specific about how one might generate a pool of appropriate indicators for the different domains. An exception is the proposed poverty-monitoring tool of Henry et al. (2000),[2] who

---

2   This tool was developed for the Consultative Group to Assist the Poor (CGAP), and was

compiled from the literature an exhaustive list of indicators of human, material, and social capital, as well as of indicators related to the fulfillment of basic needs, and then pared down the list based on eight pre-determined criteria to create a generic questionnaire that must subsequently be modified in each context. The generic questionnaire contains questions on: family size, and the age and number of children; quality of housing; type, number and value of assets, and level of school education and occupation of household members. The questions on the fulfillment of basic needs relate to hunger episodes and types of food eaten, and household expenditures on clothing.

It is interesting to note that an important body of participatory studies of poor people's own perceptions of their condition (Brock, 1999) reveals a number of prominent themes – such as difficult household relations, and violence and crime – that do not generally appear in theory-based disaggregations of socio-economic status. It is not clear whether these domains should be considered part of the definition of socio-economic status, or whether they are correlated but distinct experiences.

Domains and Indicators Included in the INDEPTH Measures of Socio-economic Status

Perhaps because the studies presented in this volume mostly built on data that were already being collected for other reasons, there is no unified conceptualization and operationalization of socio-economic status that emerges from the volume as a whole. Table 1.1 shows the various indicators that are used to characterize socio-economic status in each study. It is striking that while all studies have included multiple indicators of material capital, less than half the studies have considered human capital, and just two studies explicitly set out to measure social capital/social support. Each of these domains is considered in more depth in the following paragraphs.

All of the INDEPTH studies include information on household- or dwelling-level ownership of assets in their indices of socio-economic status. However, the number of such assets considered varies from just two (Bandim study) to twenty (Rufiji study). Non-productive assets, especially consumer durables, tend to predominate, although many studies also collected information on animals and/or vehicles.[3] Most studies simply asked whether the household owned any number of a particular kind of asset, although the Navrongo study did record the quantity of each kind of asset that the inhabitants of the dwelling owned. Morris et al. (2000) have previously shown that the value of each asset can be approximated by the relative frequency of ownership of that asset in the community, making this laborious element of data collection unnecessary.

Only one study (Farafenni) asks about monetary savings, and only two studies ask about land ownership. Obtaining accurate information about plot size can be time consuming and is sometimes seen as sensitive information; on the other hand, land is

---

intended to provide rigorous data on the poverty levels of the clients of microfinance institutions relative to other families living in the same communities.

3 As Sahn and Stifel (2003) have pointed out, there are good reasons for focusing on *productive* assets, because it is precisely the scarcity/maldistribution of these productive assets which generates poverty and income inequality.

by far the most valuable asset that many rural families in developing countries own, and not asking about it seems like a problematic omission. The majority of studies also ask about the construction of the dwelling, and the facilities present. These variables assess both the ability of the family to meet their basic needs for shelter, and also act as a proxy for the value of another prime asset in poor regions, the family house. This is especially important because improving the structure of the family dwelling is often one of the first priorities that poor people attend to as their standard of living improves.

Oakes and Rossi (2003) consider that income flows also constitute part of the domain of material capital. The FilaBavi study attempted to collect income data, although it is not possible to assess the quality of these data. One other study (AMMP) limited itself to quantifying the number of persons in salaried employment, and the main source of cash income. Obtaining full income data in addition to a range of other indicators of socio-economic status is probably beyond the capacity of most health surveys. The same applies to consumption (expenditure) data, which was, however, collected in the FilaBavi study. The CGAP tool discussed previously (Henry et al., 2000) recommends collecting data on expenditure on clothes and footwear; this sub-set of expenditure is found to be highly correlated with total expenditure in a variety of settings.

The characterization of stocks of human capital in the INDEPTH studies tends to be much more superficial than the characterization of material capital. All DSS sites collect information on education, either of the household head or of a particular individual. Only one study has actually incorporated this information into the composite index of socio-status, with the remainder analyzing differentials in health outcomes by educational level separately from the analysis of socio-economic differentials. It can be argued that education should not be included in a composite index of socio-economic status because the pathways by which it influences health outcomes are distinct and specific. This argument, however, is hard to reconcile with any approach that views measuring socio-economic status as 'uniquely locating the status of individuals in the social structure' (Oakes and Rossi, 2003), as educational attainment is a key element of an individual's location within the social structure in virtually all contemporary societies.

Human capital also includes the availability to a household of labor resources, which are often in fact the only form of capital available to poor families. Only one of the INDEPTH studies (AMMP) seeks to quantify this resource, both in terms of overall magnitude (household size) and demographic composition (summarized in the dependency ratio). Very selective indicators, such as the age and sex of the household head, are unlikely to capture the true extent of households' access to human capital, and no attempt has been made in the INDEPTH studies to determine whether any of the household members were disabled, which would obviously compromise their ability to participate in the labor market.

In the Bandim study, social capital is measured in relation to personal linkages with health professionals, and not as part of a composite index of socio-economic status. In the Farafenni study, a variety of questions are asked about parents' contact with their birth families, membership of village groups, and perceived social support from the spouse, family, and community; these variables are kept separate from indicators of socio-economic status. In general, social capital is a relatively new area of research,[4] and its importance remains somewhat controversial, making it unsurprising that it has not been widely characterized in the INDEPTH studies.

Other indicators included in composite indices in the INDEPTH studies are indicators of fulfillment of basic needs, primarily foods. Additional indicators analyzed but not included in composite indices include indicators of social status such as ethnicity and marital status, and indicators of location (geographic accessibility).

The same indicators have sometimes been analyzed at different levels in the different studies. For example, the AMMP study includes the number of individuals employed in the household in its composite socio-economic index. This distinction raises questions about the importance for health outcomes of the *intra*-household distribution of power and influence, which tend to be glossed over in analyses that only measure assets owned at the household level. The Navrongo and Farafenni studies are even less specific, since they were only able to measure assets owned in the *dwelling*, in contexts in which a dwelling might include numerous households. The Navrongo study attempts to compensate for this problem by dividing the number of assets of a particular type in a compound by the number of residents. However, this will have led to the misclassification of the socio-economic level of significant numbers of households, and is not an optimal solution. Clearly, the underlying model of all the studies is that access to resources affects individuals' health outcomes. Since the definition of a household is a group of people who pool resources, quantifying socio-economic status at the level of the household should result in meaningful and illuminating contrasts in most cases.

Methods Used to Reduce the Pool of Indicators to a Manageable Number, and to Summarize Multiple Indicators in a Single Index

In only one of the INDEPTH studies (AMMP) was a systematic method used to reduce a large number of *potential* indicators of socio-economic status to a reduced sub-set of *optimal* indicators. At least four methods are currently available to do this, and are reviewed in the following paragraphs. All of these methods yield a single, composite index, and this aspect of the analysis is therefore considered in conjunction with the problem of selecting a optimal sub-set of socio-economic indicators.

The first approach, described in detail in the AMMP study, is generally known as the 'proxy means test' approach (Grosh and Baker, 1995) because it was developed with the objective of screening potential beneficiaries of social safety net programs around the world. In this approach, a 'gold standard' measure of socio-economic status is identified, which is usually *per capita* consumption, or – less frequently – income. A model development data set is identified which includes both this gold standard measure and a full set of candidate indicators. A multiple linear regression model (Kirkwood and Sterne, 2003, Chapter 11) is then specified with the gold standard measure as the dependent (or 'outcome') variable and the candidate indicators as predictor (or 'exposure') variables. A sub-set of potential indicators mostly strongly associated with the gold standard is then identified using an appropriate model selection strategy; a 'stepwise' algorithm is commonly used, although it does

---

4   See, for example, Burt (2000).

have known disadvantages (Kirkwood and Sterne, 2003, Chapter 29). The output from this operation is a set of regression coefficients, b_k, which allow the multiple indicators to be combined in a single index (equal to b_1 times the first indicator, plus b_2 times the second indicator, plus b_3 times the third indicator, etc.). This index closely replicates the original gold standard measure, and from that point onwards, only the small subset of predictor variables needs to be collected. The main advantage of this approach is that is easily replicated and produces an index with known technical properties (see below). The main disadvantage is that the resulting index can only be theoretically justified with reference to the original gold standard, and this may itself be considered an inadequate measure of socio-economic status, as we have noted previously.

The second approach to variable selection is entirely data driven, and was popularized by Filmer and Pritchett (2001). It is based on principal components analysis (Bryman and Cramer, 2001, Chapter 11), which is a technique for extracting from a large number of variables those few linear combinations of the variables that best capture the common information. The output is a series of weights (much like the coefficients in the previous paragraph) associated with each variable, but instead of trying to mimic a gold standard, the index simply aims to maximize the amount of information common to all the variables. The critical assumption, identified by Filmer and Pritchett (2001), is that 'household long-run wealth explains the maximum variance (and covariance) in the asset variables'. This may or may not be true, and highlights the need for a strong theoretical model to guide the original variable selection. Since it is easy to establish which of the original variables are not so strongly associated with the derived composite index, these can be 'pruned away' to leave a more parsimonious subset of indicators. Such a procedure is described in Henry et al. (2000), but was not done by any of the INDEPTH studies even though the majority of them used principal components analysis to derive their indices. The main advantage of this approach is that the index should capture the latent essence of socio-economic status that is assumed to be present in each one of the multiple indicators. The main disadvantage is that the procedure is completely atheoretical. Some researchers have also objected that principal components is intended for the reduction of continuous, not categorical or binary variables, but it seems unlikely that the practical consequences of this violation of model assumptions will be very significant. The method does, however, break down when the number of indicators included in the analysis is very limited, as the resulting index becomes non-continuous and difficult to divide into equal-sized groups. A minor variation on this approach involves the use of factor analysis in preference to principal components analysis. This approach has been taken – and extensively discussed – by Sahn and Stifel (2003).

An alternative approach has been proposed by Ferguson et al. (2003). These authors set out to estimate 'permanent income', a long-run measure of income that cannot be directly observed. They assume that there exist cut-off points on the continuum of permanent income above which households are more likely than not to own a particular asset. Using a novel statistical method, they are able to locate the 'ownership thresholds' for different assets on the underlying latent variable of permanent income. This threshold information can be combined with household-level data to estimate the level of permanent income for each household. Substantial reduction of the total number of indicator variables needed to estimate permanent income is easily achieved by judiciously selecting a few assets with ownership thresholds that are evenly spaced along the permanent income continuum from high to low. The main advantage of this approach is that it is strongly grounded in permanent income theory and has potential for cross-country comparisons. The main disadvantage is that the estimation method is non-standard and demands a high level of statistical sophistication.

Finally, Oakes and Rossi (2003) have proposed a rather different method that would treat social norms as the gold standard of socio-economic status, in preference to any existing measure. 'People' (exactly who is not clear) could then be asked to assign a 'socio-economic status score' to prototypical individuals, and the revealed associations between the socio-economic status scores and the different domains represented in the prototypes would inform the final selection of weights. While highly intuitively appealing, and responding to some degree to the need to 'democratize' the definition of poverty identified in the participatory literature, this approach remains to be fully operationalized.

Of all these approaches, principal components analysis remains by far the most popular in the epidemiologic literature. Most of the INDEPTH studies reported in this volume used this method. The INDEPTH researchers eschewed *ad hoc* systems

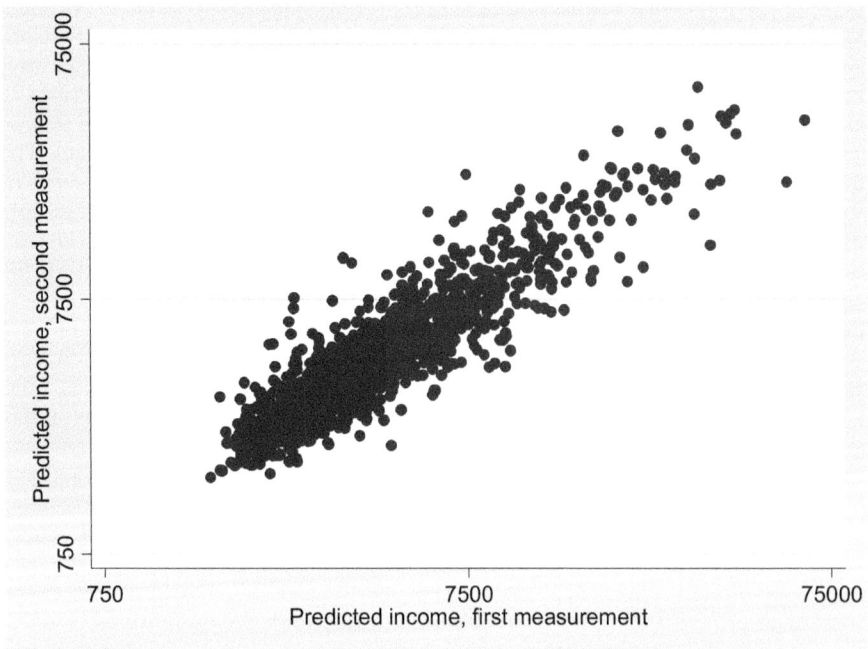

**Figure 1.1   Reliability of a proxy method for estimating income. Western
Honduras, 2000**

for combining information on various indicators into a composite index. The disadvantage of *ad hoc* systems is they cannot offer either theoretical or statistical arguments to support the content validity of the index (see below).

Reliability and Validity of Different Measures of Socio-economic Status

One of the great strengths of epidemiology is that it takes very seriously issues of the reliability and validity of different measures. These concepts are less familiar in other disciplines such as economics. Consequently, there is a fairly limited amount of documentation of the measurement properties of the different measures of socio-economic status described in previous sections. This literature is summarized in the following paragraphs.

Reliability is defined as 'the extent to which … any measuring procedure yields the same results on repeated trials' (Carmines and Zeller, 1979). Oakes and Rossi (2003) have lamented the paucity of studies of the technical properties of measures of socio-economic status, and there do not in fact appear to be any studies of the reliability of any of the measures discussed in the previous section. Figure 1.1 shows previously unpublished data from a study in Honduras, in which the proxy means test approach was used to characterize socio-economic status. In this study, 1,735 randomly selected households scattered over an area of 40 municipalities had their socio-economic status assessed on two occasions by independent, rigorously trained data collection teams. The two measurements were undertaken not less than 15 days and not more than 60 days apart, and the method used to determine the socio-economic status of these households was exactly the same on each occasion. Clearly, there is a good correlation between the two measures; on the other hand, it is far from perfect, with an intra-class correlation of 0.88 on an untransformed scale and 0.91 on a log-transformed scale.[5] It should be noted that the recall problems involved in assessing income and expenditure are much more significant than for the simple indicators used to derive the proxy means test estimator, and it is therefore likely that the reliability of the former measures is considerably lower.

There are several possible approaches to the assessment of validity. *Concurrent criterion-related validity* assesses whether a novel measure is so highly correlated to a gold standard that it could effectively replace it. Some of the evidence on this topic is summarized in Table 1.2. It is clear from this summary that there is a large variation in the performance of different measures of socio-economic status, and of the same measure across different countries and datasets. In addition, estimates of criterion-related validity do not appear to be independent of the criterion measure used. The correlations documented vary from as low as 0.18 to as high as 0.81. On average, the proxy means test indices, and the methods described in Morris et al. (2000), tend to be somewhat more highly correlated with the criterion values than do the other types of indices. This is hardly surprising, because these method had as their principal objective the maximization of this correlation. It might, in fact, be considered surprising that they did not yield even higher correlations; this is probably because the

---

5   Intra-class correlation coefficients vary from zero (indicating no more similarity in the two
    measurements than would be expected by chance alone), to unity (indicating perfect
    replication).

criterion is itself was measured with a great deal of error, as described in the previous paragraph. It is mathematically impossible for correlations between a criterion measure and a proxy for that measure to exceed the reliability of the criterion measure itself.

Unfortunately, the value of these assessments is limited by the fact that in many cases it would be difficult to argue that the criterion measures chosen are broad measures of socio-economic status. This recalls Cronbach's proposition that 'all validation reports carry the warning clause, "Insofar as the criterion is truly representative of the outcome we wish to maximize"' (Cronbach, 1971).

A second type of assessment of validity concerns *content validity*, the extent to which a measure assesses all the important aspects of phenomenon that it claims to measure. For a measure of socio-economic status to display this kind of validity, it would need a strong theoretical basis (in order to identify all relevant domains), and a systematic approach to indicator selection. Both the proxy means test approach, and that of Oakes and Rossi (2003) are particular strong in this respect, as they are built on neoclassical economic theories of household welfare, and Coleman's (1990) social theory, respectively, and offer clear guidance about indicator selection and data reduction. Ferguson et al. (2003) appeal to permanent income theory, and Henry et al. (2000) are theoretically eclectic but carefully justify all of their recommendations on domains to be included in their index and data reduction. On the other hand, asset indices are difficult to justify in terms of their content validity unless it can be argued that they provide a comprehensive assessment of wealth.

Finally, *construct validity* concerns the ability to demonstrate that other outcomes which should – theoretically – be associated with socio-economic status are associated empirically. Since there are strong theoretical reasons to believe that many health outcomes are associated with socio-economic status, much of the work in this volume can be considered to support the construct validity of the different measures employed. Whether there is convincing evidence that one type of indicator is 'more valid' (in terms of construct validity) than another for the measurement of health inequities is discussed in the next section.

How much perfection is really required in the development of measures of socio-economic status for the assessment of health inequities?

Whether or not a particular measure of socio-economic measure is 'valid' can only be answered in the context of the use to which it is being put. Wagstaff and Watanabe (2003) investigated specifically whether asset-based wealth indices (derived using principal components analysis) and household consumption were associated with the same magnitude of inequalities in child anthropometric status in 19 developing countries. They found that although on average socio-economic inequalities in malnutrition were larger by consumption than by the wealth index, the differences were very small and rarely statistically significant. This study used the concentration index (Wagstaff et al., 1991) as the preferred measure of inequality. Sahn and Stifel (2003) examined the prevalence of stunting in eleven rural developing country settings, by quintile of reported consumption and of an asset index developed using factor analysis. They did not formally assess the differences between the two approaches, but visual examination of the data suggests that there were no systematic advantages to using one approach or the other. Oakes and Rossi (2003) did not explicitly examine inequalities, but did find that, in a US database, their preferred measure of socio-economic status appeared to be more strongly associated with general health, diabetes and cardiovascular disease than was household income. On the other hand, household income was possibly more strongly associated with activities of daily living and depression. None of these differentials were formally evaluated for statistical significance.

Houweling and co-workers (2003) examined whether the magnitude of socio-economic differentials in under-5 mortality rates and measles vaccination immunization rates in ten developing countries was affected by the choice and number of socio-economic indicator variables included in a summary index derived using principal components analysis. They confirmed that trying to base the index on small numbers of indicator variables often led to heaping on relatively few values, such that it was impossible to divide the sample into five equally sized groups. Furthermore, the magnitude of inequalities documented was sensitive to the subset of indicators variables included, often substantially so, with the relative index of inequality changing by more than 30%.

It would seem that no studies of health outcomes have compared the more complex measures of socio-economic status with simpler ones such as unweighted sums of different kinds of assets owned. One study has made these comparisons focusing on fertility as an outcome rather than health (Bollen, Glanville and Stecklov, 2001). These authors found that, of a total of ten different measures of socio.economic status, an asset index derived using principal components and a 15-17 different indicators provided the best fit to the data. In this study, two different datasets, from Ghana and Peru, gave virtually identical rankings of the ten different measures. Incorporating asset values into the index was found to *decrease* the explanatory power of the index, as was reducing the number of assets considered.

Conclusion

Documenting the magnitude of socio-economic inequities in health outcomes is a critical first step in the design of appropriate programmatic responses. In doing this, the development of an appropriate household-level index of socio-economic status is aided by a rapidly growing body of literature which provides both the necessary theoretical framework and practical guidance on statistical analysis. The studies presented in this volume illustrate that these methods are not inaccessible to field-based research institutions in developing countries. In general, experience has shown that it is possible to replace the complex and time-consuming measures favored by micro-economists with approaches that are considerably better suited to large-scale epidemiologic survey work. On the other hand, exaggerated reduction of the number of indicator variables used to derive these indices is likely to compromise their validity, and ignoring conceptual issues in their development jeopardizes the meaningfulness of the comparisons that can be made.

References

Bollen, K.A., Glanville, J.L. and Stecklov, G. 'Economic status proxies in studies of fertility in developing countries: does the measure matter?' MEASURE Evaluation working paper 01-38. Chapel Hill, NC: MEASURE Evaluation, 2001.

Brock, K. 'It's not only wealth that matters – it's peace of mind too': A review of participatory work on poverty and ill-being. Brighton, UK: Institute of Development Studies, 1999.

Bryman, A. and Cramer, D. *Quantitative data analysis with SPSS for Windows: A guide for social scientists* (Release 10). London: Routledge, 2001.

Burt, R. 'The network structure of social capital'. In: Staw, B.M. and Sutton, R.L. (eds). Research in organizational behaviour, Volume 22. Greenwich, CT: JAI Press, 2000.

Carmines, E.G. and Zeller, R.A. 'Reliability and validity assessment'. Sage university paper series on quantitative applications in the social sciences, 07-017. Newbury Park, CA: Sage, 1979.

Coleman, J.S. *Foundations of social theory.* Cambridge, MA: Harvard University Press, 1994.

Cronbach, L.J. 'Test validation'. In: Thorndike, R.L. (ed.). *Educational measurement.* Washington, DC: American Council on Education, 1971.

Deaton, A. *The analysis of household surveys: a microeconometric approach to development policy.* Washington, DC: The World Bank, 1997.

Falkingham, J. and Namazie, C. *Measuring health and poverty: a review of approaches to identifying the poor.* London, UK: DfID Health Systems Resource Centre, 2002.

Ferguson, B., Murray, C., Tandon, A. and Gakidou, E. 'Estimating permanent income using asset and indicator variables'. Evidence and Information for Policy Discussion Paper No. 44. Geneva: World Health Organization, 2000.

Filmer, D. and Pritchett, L.H. 'Estimating wealth effects without expenditure data—or tears: an application to educational enrollments in states of India'. Demography 2001; 38(1): 115-32.

Grosh, M.E. and Baker, J.L. 'Proxy means tests for targeting social programs: simulations and speculation'. Living Standards Measurement Study working paper no. 118. Washington, DC: World Bank, 1995.

Henry, C., Sharma, M., Lapenu, C. and Zeller, M. *Assessing the relative poverty of microfinance clients: a CGAP operational tool.* Washington, DC: International Food Policy Research Institute, 2000.

Houweling, T.A.J., Kunst, A.E. and Mackenbach, J.P. 'Measuring health inequality among children in developing countries: does the choice of indicator of economic status matter?' Int J Equity Health 2003; 2: 8.

Kasl, S.V. and Jones, B.A. 'Social epidemiology: towards a better understanding of the field'. Int J Epidemiol 2002; 31: 1094-7.

Kawachi, I. 'Social epidemiology'. Soc Science Med 2002; 54: 1739-41.

Kirkwood, B.R. and Sterne, J.A.C. *Essential medical statistics.* 2nd edition. Malden MA: Blackwell Science Ltd., 2003.

Krieger, N. 'Sticky webs, hungry spiders, buzzing flies, and fractal metaphors: on the misleading juxtaposition of "risk factor" versus "social" epidemiology'. J Epidemiol Community Health 1999; 53: 678-80.

Krieger, N. 'Epidemiology and social sciences: towards a critical reengagement in the 21st century'. Epidemiol Rev 2000; 22(1): 155-63.

Krieger, N., Williams, D.R. and Moss, N.E. 'Measuring social class in US public health research'. Ann Rev Public Health 1997; 18: 341-78.

Johnson, M., McKay, A.D. and Round, J.I. 'Income and expenditure in a system of household accounts: concepts and estimation. Social dimensions of adjustment in sub-Saharan Africa'. Working paper no. 10. Washington, DC: World Bank, 1990.

Morris, S.S., Carletto, C., Hoddinott, J. and Christiaensen, L.J. 'Validity of rapid estimates of household wealth and income for health surveys in rural Africa'. J Epidemiol Community Health 2000; 54(5): 381-7.

Oakes, J.M. and Rossi, P.H. 'The measurement of SES in health research: current practice and steps towards a new approach'. Soc Science Med 2003; 56: 769-84.

Pan-American Health Organization (PAHO). 'Introduction to social epidemiology'. Epidemiol Bull 2002; 23(1): 7.

Sahn, D.E. and Stifel, D. 'Exploring alternative measures of welfare in the absence of expenditure data'. Rev Income Wealth 2003; 49(4): 463-89.

Wagstaff, A., Paci, P. and van Doorslaer, E. 'On the measurement of inequalities in health'. Soc Science Med 1991; 33: 545-557.

Wagstaff, A. and Watanabe, N. 'What difference does the choice of SES make in health inequality measurement?' Health Econ 2003; 12(10): 885-90.

Table 1.1   Indicators used to assess socio-economic status in the INDEPTH studies

Site	Material capital	Human capital	Other
Agincourt	Appliance, vehicle, and livestock ownership; Construction of the dwelling; Water and electric connections.		
AMMP	Land area; Iron, stove, car, bicycle, sofa, lamp ownership; Wall construction; Dwelling size; Source of drinking water; Toilet facility; Fuel used for lighting No. persons employed and main source of cash income.	Household size; Education of household head; Age of household head; Sex of household head; Dependency ratio.	**Indicators of fulfillment of basic needs:** No. days meat, milk products eaten/week; Expenditures on particular foods, seeds, fertilizers;
Bandim	Television ownership; Construction of roof; Electric connection to the dwelling; Dwelling size; Toilet facility.	Maternal education.*	Beliefs, as assessed by agreement with pre-defined statements;* Personality type as assessed by interviewer;* Self-reported social capital.* Ethnicity.*
Farafenni	Bicycle, radio, bed, cattle and sheep/goat ownership; Construction of roof; Sales of groundnuts; Possession of money; Possession of goods to be sold; Possession that could be used as collateral.		**Indicators of fulfillment of basic needs:** Weekly consumption cooking oil. **Indicators of social support:** Contact with birth family;* Group membership;* Perceived support.*
FilaBavi	Land area; Household assets; Housing conditions; Sanitary conditions; Income; Expenditures.		

Table 1.1 continued

Site	Material capital	Human capital	Other
Ifakara	Bicycle, radio, poultry, tin roof, livestock ownership; Housing tenure.		
Matlab	Ownership of consumer durables; Construction of roof and wall; Source of drinking water; Toilet facility.		
Navrongo	No.s of bicycles, beds, donkey carts, radios, sewing machines, lamps, coal pots, cows, sheep, goats, and pigs owned; Dwelling design (modern); Construction of roof; Dwelling size; Source of water.	Education of dwelling head*, mother;* Mother's age.*	**Social status:** Ethnicity;* Mother's marital status.* **Environment:** Distance to health facility;* Type of health service delivery strategy.*
Rufiji	Ownership of 22 assets; Housing tenure; Construction of roof, wall, and floors; Dwelling size; Electric connection to the dwelling; Source of drinking water, and distance; Fuel used for cooking; Toilet facility.		
Watch	Land ownership;* Table, cot, quilt, watch, radio, television, and bicycle ownership; Electric connection to the dwelling.	Individual's age;* Individual's educational level;*	**Environment:** Region of country.*

* Information analyzed, but not combined with other indicators in a single index.

Table 1.2   Criterion-related validity of different measures of socio-economic status, as documented in published studies

Author and publication	Location	'Test' measure of socio-economic status	Criterion	Correlation
Grosh and Baker, 1995	Jamaica	Proxy means test index	(Log) per capita consumption	0.64
	Urban Bolivia	Proxy means test index	(Log) per capita consumption	0.59
	Lima, Peru	Proxy means test index	(Log) per capita consumption	0.56
Grosh and Glinskaya, 1998	Armenia	Proxy means test index	(Log) per capita consumption	0.56
Ahmed and Bouis, 2001	Egypt	Proxy means test index	(Log) per capita consumption	0.66
PRAF/IFPRI, 2000	Rural Honduras	Proxy means test index	(Log) per capita consumption	0.79
IFPRI, 2002	Rural Nicaragua	Proxy means test index	(Log) per capita consumption	0.71
Skoufias and Coady, 2000	Rural Mexico	Proxy means test index	(Log) per capita consumption	0.78
Setel et al., *this volume*	Rural Tanzania [1]	Proxy means test index	(Log) per capita consumption	0.81
	Rural Tanzania [2]	Proxy means test index	(Log) per capita consumption	0.75
	Dar es Salaam, Tanzania	Proxy means test index	(Log) per capita consumption	0.79
Sahn and Stifel, 2003	Cote d'Ivoire	Asset index by factor analysis	(Log) per capita consumption	0.51*
	Ghana [1]	Asset index by factor analysis	(Log) per capita consumption	0.43*
	Ghana [2]	Asset index by factor analysis	(Log) per capita consumption	0.42*
	Jamaica	Asset index by factor analysis	(Log) per capita consumption	0.39*
	Madagascar	Asset index by factor analysis	(Log) per capita consumption	0.50*
	Nepal	Asset index by factor analysis	(Log) per capita consumption	0.55*
	Pakistan	Asset index by factor analysis	(Log) per capita consumption	0.42*
	Papua New Guinea	Asset index by factor analysis	(Log) per capita consumption	0.47*
	Peru	Asset index by factor analysis	(Log) per capita consumption	0.71*

Table 1.2 continued

Author and publication	Location	'Test' measure of socio-economic status	Criterion	Correlation
Sahn and Stifel, 2003, cont.	South Africa	Asset index by factor analysis	(Log) per capita consumption	0.71*
	Viet Nam [1]	Asset index by factor analysis	(Log) per capita consumption	0.55*
	Viet Nam [2]	Asset index by factor analysis	(Log) per capita consumption	0.67*
Ferguson et al., 2003	Greece	Latent variable estimation using a modified hierarchical probit model	Total household income	0.60
			Per capita income	0.41
			Income per adult equivalent	0.56
	Pakistan	Latent variable estimation using a modified hierarchical probit model	Total household income	0.17
			Total household consumption	0.33
			Income per capita	0.18
			Income per adult equivalent	0.18
			Per capita consumption	0.30
			Consumption per adult equivalent	0.34
	Peru	Latent variable estimation using a modified hierarchical probit model	Total household income	0.59
			Total household consumption	0.61
			Income per capita	0.52
			Income per adult equivalent	0.58
			Per capita consumption	0.48
			Consumption per adult equivalent	0.59
Oakes and Rossi, 2003	USA	Composite index with principal components analysis used to weight items in each of three domains	Household income	0.52

Table 1.2 continued

Author and publication	Location	'Test' measure of socio-economic status	Criterion	Correlation
Morris et al., 2000	Mali	Asset ownership index, weighted by frequency of ownership in community	(Log) value household assets	0.74
	Malawi	Asset ownership index, weighted by frequency of ownership in community	(Log) value household assets	0.83
	Côte d'Ivoire	(Log) household consumption on 10 selected items	(Log) total household consumption	0.72-0.74

* Rank correlation.

Chapter 2

Socio-economic Status and Health Inequalities in Rural Tanzania: Evidence from the Rufiji Demographic Surveillance System

Eleuther Mwageni, Honorati Masanja, Zaharani Juma,
Devota Momburi, Yahya Mkilindi, Conrad Mbuya, Harun Kasale,
Graham Reid and Don de Savigny

Summary

It has been questioned whether demographic surveillance system (DSS) sites, which normally operate in relatively small, homogeneous areas, and in relatively small populations of the order of 100,000 people, are large enough to examine inequalities and inequities in health. The Rufiji DSS in Tanzania has attempted to apply principal components analysis (PCA) to asset and other household data collected in the routine course of a DSS, to rank individuals according to a household socio-economic index and investigate whether this predicts health system access or outcomes.

In this study, we determined wealth indices for individuals in 12,604 rural households in the Rufiji DSS area using principal components analysis for the year 2000. The index was based on the presence or absence of items from a list of 20 specific household assets and 9 household characteristics dealing with household ownership, construction features, water supply, sanitation, and type of fuel. PCA revealed 49 principal components of which the first accounted for 12.9% of the total variance. Asset ownership and housing features contributed equally to the variance in the first component. Scores for each asset or feature were internally consistent with expectations. In this study, we examined mosquito net ownership as an example of health intervention access, and infant, child and under-5 mortality rates as examples of health outcomes, all in relation to quintiles of populations and determined poorest-least poor ratios, concentration indices, and chi-square for linear trend.

Results showed significant gradients in both access and outcome measures across wealth quintiles, even in this relatively homogeneous rural area. Poorest-least poor ratios for infant, child and under-5 mortality were 1.46, 1.41 and 1.53 respectively while lack of access to mosquito nets was 1.82. The findings call for more attention to strategies or approaches for reducing health inequalities. These could include reforms in the health sector to provide more equitable resource allocation, improvement in the quality of the health services offered to the poor, and redesigning interventions and their delivery to ensure they are more pro-poor. Such proactive measures will be important if health equity goals at community level are to be achieved.

Background

Efforts to improve health in developing countries face many challenges. These include high incidence of infectious and communicable diseases, growing burdens of chronic and non-communicable diseases, weak health systems, and inadequate human and material resources. There are also unquantified and poorly understood socio-economic inequalities in access to health services within and between various population groups. Little is known of the factors that determine these inequalities and the mechanisms through which they operate in various sub-groups.

The relationship between socio-economic differentials and health status in developing countries has been documented in several studies (Caldwell, 1979; Cochrane, et al. 1982; Rutstein, 1984; Bicego and Boerma, 1993; Gwatkin, et al., 2000; Woelk, 2000). Using a study of 20 cross-sectional Demographic and Health Surveys (DHS) from developing countries Bicego and Ahmad (1996) found that mortality risks of under-5s born to uneducated women were more than twice as high as to those born to women with a secondary education. Gwatkin et al. (2000) using DHS data from Tanzania described differences between the poor and the least poor in mortality, nutrition and treatment of illnesses. Woelk and Chikuse (2000) in Zimbabwe showed that stunting, underweight and occurrence of diarrhoea varied according to the socio-economic status noting that being in the lowest socio-economic status increased the risk of being underweight for children by about three times compared to those in the highest socio-economic group. Filmer (2002) in a study of 22 malarious countries in Africa found little difference in rates and risk of fever among the poor and least poor, but the poorest had a much smaller likelihood of obtaining suitable treatment. Armstrong Schellenberg et al. (2003) made similar observations of poor/least poor inequities across a broader array of childhood illnesses and health interventions using principal components analysis of household data from a cross-sectional survey in rural Tanzania.

The purpose of the work reported here is to provide a description of socio-economic differentials in relation to health status and health service access by using data from longitudinal demographic surveillance survey systems (DSS). The major objective of the study was to explore the utility of DSS as a practical and continuing source of equity monitoring data. This entailed conducting asset and housing condition surveys nested into DSS surveillance and using principal components analysis to build an asset and housing based socio-economic status index. This was used to determine the relationship between household socio-economic characteristics and inequalities of access to health interventions, and to health outcomes in rural Tanzania. Specifically this study examined how proxies for socio-economic status (e.g. ownership of assets, housing quality and sanitation) relate to mortality in infants and children under five years old mortality as well as ownership of mosquito nets, a preventive health intervention. The data obtained is intended to assist programme providers and policy makers to recognize health system inequities in order to improve the health status in rural areas.

Methods

The Rufiji Demographic Surveillance System (DSS) commenced field operations in November 1998. The DSS approach involves continuous monitoring of households and members within households in cycles or intervals, known in the Rufiji DSS as 'rounds' of four months each. The Rufiji DSS collects information on demographic, household, socio-economic and environmental characteristics of a population of about 87,000 people in 31 villages located in Rufiji District along the coastal area of Tanzania, south of Dar es Salaam in the Rufiji River basin. The Rufiji DSS was established as one of the four major research components of the Tanzania Essential Health Interventions Project (TEHIP). In addition to its research role, its aim is to provide sentinel data to the district health authorities and the Ministry of Health to inform evidence based planning and resource allocation as well as to quantify the burden of disease and document impact of health system interventions and innovations.

Rufiji is one of the six districts of Coast region and has an estimated population of about 187,000. The Rufiji River running West to East cuts the district roughly into half. The surveillance area is a contiguous area situated on the northern side of the river floodplain. The Rufiji DSS employs the Household Registration System (HRS) (Indome et al., 1995), which involves collecting and documenting data on pregnancies and births, deaths, causes of death, in and out-migrations and socio-economic status. Full details on the Rufiji DSS have been reported elsewhere (Mwageni et al., 2002).

The data for this study come from routine core demographic information, mortality data and socio-economic data collected by the Rufiji DSS for the years 1999 and 2000. Specifically for this report, data comes from 16,260 active households. Of these 12,604 households (about 77.5%) had consistent data for assets, housing conditions, water and sanitation variables sufficient to create a household socio-economic status index. Socio-economic data was collected during the October 2000-January 2001 Rufiji DSS round coinciding with the end of the two-year mortality data set. The data collected included: asset ownership, housing conditions, source of energy for cooking, and type of water and sanitation. The socio-economic questionnaire was developed through reviewing multiple sources such as common assets owned by the community, standard lists of assets from previous studies within the country and from multi-country studies such as the DHS and those done by World Bank (See Annex 2.1).

Routine Rufiji DSS data has been analysed using HRS software (Phillips and MacLeod, 1995). Principal Components Analysis (PCA) using Stata 7.0 (Stata Corporation) was applied to the socio-economic data to obtain an index as a proxy for household socio-economic status.

PCA involves breaking down assets (e.g. radio, wrist watch) or household service access (e.g. water, electricity) into categorical or interval variables. The variables are then processed in order to obtain weights and principal components. The results obtained from the first principal component (explaining the most variability) are usually used to develop an index based on the formula:

$$A_j = f_1 \times (a_{ji} - a_1)/(S_1) + \ldots + f_N \times (fa_{jN} - a_N)/(s_N) \text{ (Filmer and Pritchett, 1998)}$$

where f_1 is the scoring factor or weights for the first asset (or service), x is the variable (asset or service), a_j is the value for the assets (or service), and a_1 and s_1 are the mean and standard deviation of assets (or service) respectively. Based on this equation socio-economic statuses of households were assigned to the residents of those households, and the resulting population was divided into quintiles that then represent proxies for socio-economic status. The quintiles developed are thus expressed in terms of quintiles of individuals of the total population at risk for all measures. The 1st, 2nd, 3rd, 4th and 5th quintiles were assigned in the continuum of poorest and least poor respectively.

Three statistical indicators of inequality were measured. One was the poorest / least poor ratio which is the ratio comparing the rate prevailing in the poorest quintile with the rate in the least poor quintile. This method ignores information contained in the middle three quintiles. The second measure used was the concentration index calculated by the method of Kakwani et al. (1998). This measures the extent to which a variable is distributed unequally across all five socio-economic quintiles, i.e. the concentration of inequality. The closer the index is to zero, the less concentrated the distribution of inequality (Gwatkin et al., 2000). The third was a trend test (Chi Square) to determine the significance of any gradient in the inequality.

Results and Discussion

Socio-economic Status Index

The final index was based on household assets, housing quality, water and sanitation and constituted the independent variable. The asset index approach has been used and recommended by many studies (Filmer and Pritchett, 1998; Bonilla-Chacin and Hammer, 1999; Wagstaff and Watanbe, 1999; Gwatkin et al., 2000; Sahn and Stiffel, 2000). In a study conducted in several states of India, Filmer and Pritchettt (1998) found that the asset index produces comparable results with other measures. The authors noted that the asset index significantly correlated with the state headcount index as well as the domestic product per capita distributions.

Complete results of the PCA are summarised in Figure 2.1. There are 49 principal components and the first component accounts for 12.9% of the total variance. The second largest component explains 5.0% of total variance of all the variables. The Eigenvectors of the first component have been used as scoring weights for each of the asset and service items. The results reveal that variance in the first component is explained by asset ownership (39.8%) and housing conditions (39.0%). Source of energy explains 14.5% of the variance while lastly water and sanitation has the least variance at 6.6%.

The asset or item scores are then used to assign a wealth index value to each household and its members. Eventually households and their members are assigned into quintiles based on the value of the asset index. For the purpose of this analysis the lowest quintile is considered as a socio-economic status (wealth) proxy for the poorest and the highest quintile represents the least poor households.

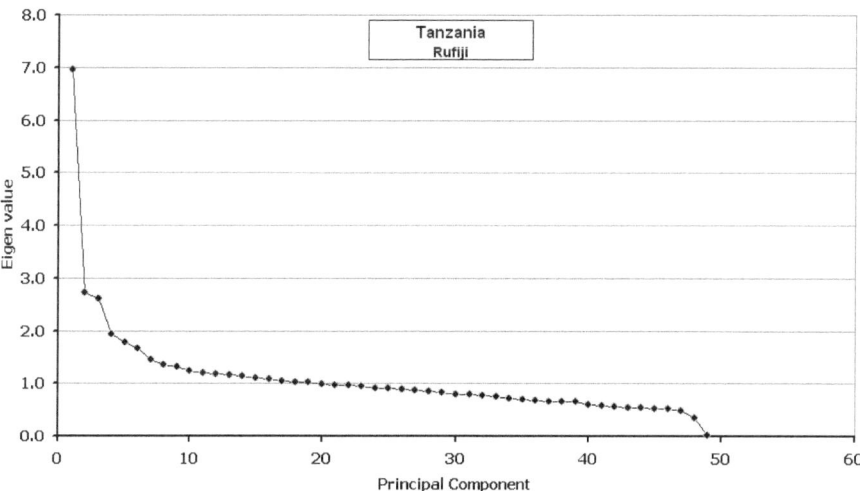

Figure 2.1   Scree plot of principal components and Eigen values

Distribution of Index Components by Socio-economic Status

The proportion of households possessing a given characteristic according to the socio-economic status of that household reveals additional interesting results (See Annex 2.2). In general the poorest are below average in most of the items or services to which the better off have access. For example, in terms of asset ownership 12% of the poorest have a bicycle compared with 55% of the least poor, four times more. The same applies for radio, sofa, mattress and wardrobes. Thus, as expected, the better off are likely to own more assets than the poorest. The exception is for land and poultry where the poor have more than the better off. These observations are consistent with the directions of the scores. Like asset ownership, housing conditions tend to reflect the economic status of the household. A similar pattern is noted for the sources of energy for cooking and sanitation. Households that ranked lower in the index are more likely than the better off to use firewood and water from public well. The congruence between the socio-economic status index and variables from which it was generated provides evidence of internal consistency of the index developed.

Socio-economic status and health

The relationship between socio-economic status and health has been an area of increasing interest (Gwatkin et al., 2000; Koenig et al., 2000; Filmer, 2002; Armstrong Schellenberg et al., 2003). The health status indicators used by this study were mortality of children under five years of age (infant, child and under-5 mortality) and the health intervention indicator was household ownership of mosquito nets. Examining the relationship between the index proposed and these indicators serves as a way of testing the consistency of the index with other data that are possibly related with socio-economic status. Differentials in health related indicators according to socio-economic groupings would imply that the index is sensitive to differences.

Socio-economic status and Health Outcomes: Mortality of children.

About 2,099 deaths were registered in the study area between 1999 and 2000, of these 705 deaths were for children below five. Table 2.1 shows data on how infant mortality is distributed across the different socio-economic groups. The findings show a significant inverse trend such that infant mortality rate declines with increase in the socio-economic status of the household. Children in the poorest households are about 46% more likely to die in infancy than those in the least poor or better off. If the socio-economic status of the poorest households were improved to the level of the better off, then about 34 lives per 1,000 infants could be saved annually (rate difference).

Table 2.1   Infant mortality by socio-economic status

Quintile	Infant Person-Years Observed (PYOs)	Infant Deaths	Infant Mortality Rate/1000 PYOs (95% CI)
1st (Poorest)	830.6	89	107.1 (87.0, 131.9)*
2nd	830.4	86	103.6 (83.8, 127.9)
3rd	839.3	69	82.2 (64.9, 104.1)
4th	824.2	68	82.5 (65.0, 104.6)
5th (Least Poor)	856.5	63	73.6 (57.5, 94.2)
Poorest–Least Poor Ratio			1.46
Concentration Index			−0.08
Chi-Square Trend			p = 0.009

* Quintiles are based on infant person-years, not households

The relationship between socio-economic status and child mortality (deaths to children between age one and four years) is presented in Table 2.2. As in the relationship shown in the infant mortality, socio-economic status has some association with child mortality, with the poorest households having higher probabilities of child death than the least poor. But contrary to the relationship shown to the infant deaths, the pattern is not consistent between the third to the fourth quintiles and there is no statistically significant trend. Those in the second quintile have lower mortality than those in the third or fourth. The reasons for this inconsistency are not known but may be due to differences in the heterogeneity of scores within quintiles. In a study that used DHS data in Tanzania an inconsistent pattern was also noted between health, nutrition, and population indicators for the second to fourth quintiles (Gwatkin et al., 2000).

Table 2.2   Child mortality by socio-economic status

Quintile	Person Years Observed	Deaths 1-4 yr	Child Mortality Rate
1st (Poorest)	2596.3	37	14.3 (10.3, 19.7) *
2nd	2612.1	22	8.4 (5.5, 12.8)
3rd	2534.7	28	11.0 (7.6, 15.9)
4th	2580.7	39	15.1 (11.0, 20.7)
5th (Least Poor)	2594.7	20	7.7 (5.0, 11.9)
Poorest–Least Poor Ratio			1.85
Concentration Index			−0.04
Chi-Square Trend			p = 0.334

* Quintiles are based on child 1-4 years person-years lived, not households

The relationship between socio-economic status and overall under-5 mortality is summarised in Table 2.3 and Figure 2.2. The data presented indicate that under-5 mortality is higher in the poorest quintiles and lower for the rest of the quintiles. Its pattern is very similar to that of the infant. The data reveal that children of the poorest are 53% more likely to die before reaching their fifth birthday than those of the better off households. The gradient for under-5 mortality is not very consistent between the fourth and fifth quintiles. The under-5s in poorest households have

Table 2.3   Under-5 mortality and socio-economic status

Quintile	Under-5 Person Years Observed	Deaths 0-4 yr	Under-5 Mortality Rate (95%)
1st (Poorest)	3405.2	125	36.7 (30.8, 43.7) *
2nd	3433.2	108	31.5 (26.1, 38.0)
3rd	3379.5	98	29.0 (23.8, 35.3)
4th	3423.9	107	31.3 (25.9, 37.8)
5th (Least Poor)	3457.8	83	24.0 (19.4, 29.8)
Poorest–Least poor Ratio			1.53
Concentration Index			−0.07
Chi-Square Trend			p = 0.007

* Quintiles are based on under-5 person-years lived, not households

similar inequitable poor-least poor risks of dying as the infants. This indicates the differentials noted at infancy have shaped the relationship between socio-economic status and under-5 mortality. If the socio-economic status of the poorest households were improved to the level of the better off, then about 13 lives per 1,000 children under five could be saved annually (rate difference).

Socio-economic Status and Mosquito Net Ownership

Consistent use of insecticide treated mosquito nets has been shown to protect individuals from insect bites, especially mosquitoes, to reduce the transmission of mosquito related diseases such as malaria in children, and to substantially lower the risk of both malaria morbidity and all-cause under-5 mortality (Lengeler, 1998). In Tanzania and in the study area, mosquito nets are not available free of charge. They are sold at either commercial or social market prices. As there is a cost component involved in the purchase of mosquito nets, the most socially disadvantaged groups may not have as much access to the protective effect of mosquito nets. The Rufiji DSS conducted a net owner and user study in the October 2000-January 2001 round during the general asset survey. The objective was to measure the coverage of mosquito net and treated mosquito net use within households especially among children under five years of age.

The relationship between socio-economic status and mosquito net ownership is presented in Table 2.4 and Figure 2.3. The results indicate consistently that non-ownership of mosquito nets is associated with the lower socio-economic quintiles. While in the poorest household only about 6% own nets, the proportion is more than

Table 2.4   Mosquito net ownership by socio-economic status

Quintile	No. Without Mosquito nets	No. of Households per Quintile	% Households in Quintile	% of Quintile Population without Mosquito nets (95% CI)
5th (Least Poor)	2723	2888	20%	94.3
4th	2470	2888	20%	85.5
3rd	2277	2888	20%	78.8
2nd	1496	2887	20%	51.8
1st (Poorest)	2277	2888	20%	78.8
			Poorest–Least Poor Ratio	1.82
			Concentration Index	−0.33
			Chi-square Trend	p<0.000

* Quintiles are based on households

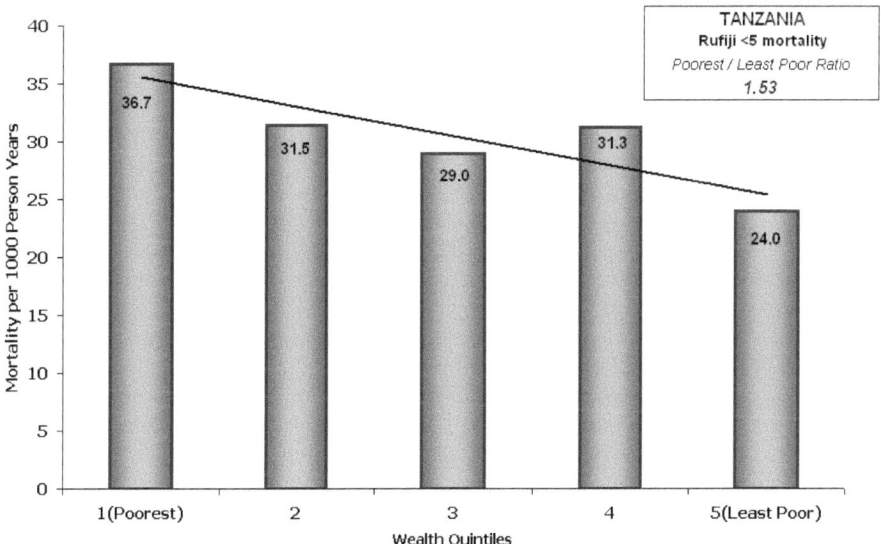

Figure 2.2   Under-5 mortality by wealth quintile in the Rufiji DSS area 2000
Note that mortality is expressed as a rate per 1,000 under-5 person years observed, and not the conventional deaths per 1000 live births or 5q0 which would be approximately 4.5 times higher.

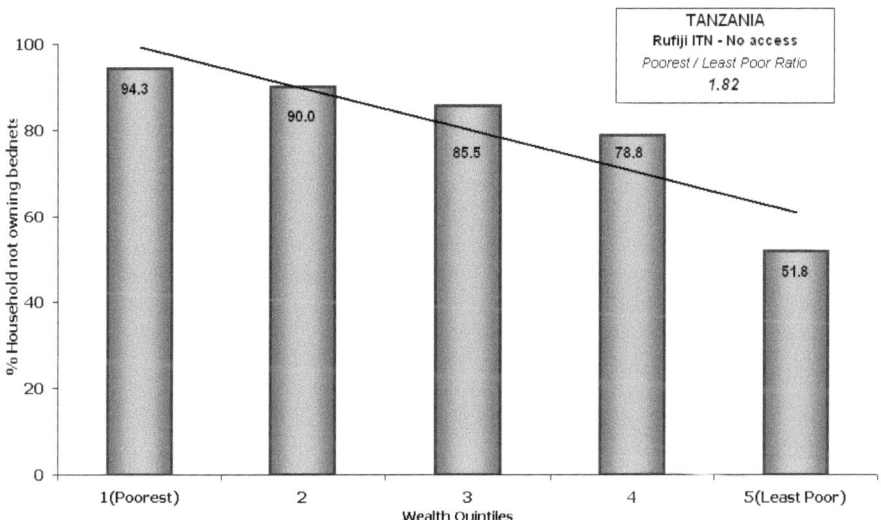

Figure 2.3   Mosquito net ownership (expressed as no coverage) in the Rufiji DSS area by wealth quintiles, 2000

eight times higher for the better off. The poorest/least poor ratio of non-ownership of 1.82 reveals that there is a large inequality between the poorest and the better off in terms of mosquito net ownership in the study area. Since the mosquito nets are sold one can postulate that the poorest are unable to purchase them at the prices offered in 2000 (approximately $3.00-4.00 USD) possibly due to low purchasing power coupled with low access to information. This result is about the same as the poorest-least poor ratio of non-coverage 1.82 found by a district-wide, randomised household survey of health behaviours conducted in Rufiji District in the previous year, 1999 (Armstrong Schellenberg, 2000) suggesting that inequalities are slowly being addressed at current prices and accelerated progress is urgently required before this ratio can approach 1.0.

Evidence however is mounting that households without nets, if in proximity (less than 300 meters) from a household with treated nets, will benefit significantly with regard to lower malaria indices, anaemia, and risk of death (Binka, Indome and Smith, 1998; Howard et al., 2000; Hawley et al., 2003). Further studies are underway in the Rufiji DSS area to determine what level of coverage, and what spatial distribution of users and non-users, will contribute to wider effective coverage of the poorest as well as the least poor.

Conclusions

This study has shown a relationship between socio-economic status and health indicators with particular focus on the differentials between the poorest and the least poor. PCA was applied to a set of asset and household variables that have relationship with socio-economic status. The first principal component, accounting for most of the variance among the asset and service variables was employed to obtain an index as a proxy of socio-economic status of the households. Based on the value of the asset and household variables as well as using the scoring weights obtained for these variables each household and its members was assigned to a specific quintile.

The study also attempted to check the internal consistency of the index developed by examining its distribution against the quintiles of the household variables that had been used for its creation. The results revealed expected patterns on how the asset and household variables change with the quintiles. This was noted for variables such as bicycles, radio, sofa, wardrobes, use of firewood and access to water. The exception is for land, poultry and owning a house where the poorest have more access than the better off. Most of these assets, being a rural area, are likely to be owned by the indigenous and not the new comers, most of whom are salaried employees. In general, the index developed appears to be useful in capturing some form of material well-being at household level.

Using quintiles generated from the PCA, the study has shown that the poorest have higher mortality rates then the least poor. In addition, the study has revealed that mosquito net ownership is wholly inadequate in the poorest households and that there are profound inequalities in access to this particular protection against malaria morbidity and mortality as delivered in the Rufiji District at the time of the survey.

The findings call for more attention to strategies or approaches for reducing health inequalities. These could include reforms in the health sector to provide more equitable resource allocation, improvement in the quality of the health services offered to the poor, and redesigning interventions and their delivery to ensure they are more pro-poor. Such proactive measures will be important if health equity goals at community level are to be achieved.

Finally, the study shows that DSS operations can host manageable asset surveys and that a PCA approach to such data is surprisingly sensitive to differences in socio-economic status. These gradients are sufficient to predict differences in health outcomes such as child mortality, as well as access to health interventions, even though the source population might appear to be broadly homogeneous with regards to poverty.

This study sets the stage for more in-depth equity research and intervention. The next stage of equity work in the Rufiji DSS will attempt to understand the determinants of these inequalities and will start by conducting a spatial analysis of the distribution of households across quintiles, and their physical access to health services. Global satellite positioning of all households has now been completed and staff have been trained in geographic information systems analysis. Inter-sectoral studies are being developed to extend the range of determinants under study (e.g. food insecurity). The long-term aim is to develop a platform for longitudinal, inter-sectoral studies to support and monitor progress in poverty alleviation and health development interventions. In parallel with these studies, work will begin on establishing routine monitoring of trends in poverty and inequality with a focus on how to package such results in ways that facilitate mitigating interventions by policy makers and service providers.

Acknowledgements

We are grateful to Dr. Davidson Gwatkin and Dr. Fred Binka for proposing and encouraging this study, and to the INDEPTH Network and the sponsors of its equity research program, the Rockefeller Foundation and the World Bank for financial support. We thank Dr. Philip Setel and the Adult Morbidity and Mortality Project team for advice and support in designing the asset index and survey. We are most grateful to the field and data staff of the Rufiji DSS who collected the socio-economic data on which this preliminary analysis has been based. The Rufiji DSS was largely supported by the International Development Research Centre (IDRC), Canada, with additional funding from the Department for International Development (DFID), UK and managed by the Tanzania Essential Health Interventions Project (TEHIP).

References

Armstrong Schellenberg, J (2000). *Personal Communication.*
Armstrong Schellenberg, J., C.G. Victora, A. Mushi, D. de Savigny, D. Schellenberg, H. Msinda and J. Bryce (2003). 'Inequities among the very poor: health care for children in rural southern Tanzania'. *Lancet* 361: 561-566.
Bicego, G. and O.B. Ahmad (1996). *Infant and Child Mortality* Demographic and Health Surveys Comparative Studies No. 20. Calverton, MD: Macro International, 1996.
Bicego, G. and J.T. Boerma (1993). 'Maternal education and child survival: a comparative study data from 17 countries'. *Social Science and Medicine* 36 (9): 1207-1227.

Bonilla-Chacin, Maria and Jeffrey S. Hammer (1999). *Life and Death among the Poorest*. The Johns Hopkins University and Development Economics Research Group, The World Bank.

Binka, F., F. Indome and T. Smith (1998). 'Impact of spatial distribution of permethrin-impregnated mosquito nets on child mortality in rural northern Ghana'. *Am. J. Trop. Med. Hyg.* 59: 80-85.

Caldwell, J.C. (1979). 'Education as a factor in mortality decline: an examination of Nigerian data'. *Population Studies* 33 (3): 395-413.

Cochrane, S.H., J. Leslie and D.J. O'Hara (1982). 'Parental education and child health: intra-country evidence'. *Health Policy and Education* 2:1330 39.

Filmer, D. and L. Pritchett (1998). 'Estimating wealth effects without income of expenditure data – or tears: An Application to Educational Enrolments in States of India'. World Bank Policy Research Working Paper.

Filmer, D. (2002). *Fever and its Treatment among the more and less poor in Sub-Saharan Africa*, World Bank Development Economics Research Group Working Paper 2789, March 2002. http://econ.worldbank.org/files/13157_wps2798.pdf

Gwatkin, D.R, S. Rutstein, K. Johnson, R.P. Pande, and A. Wagstaff (2000). *Socio-economic Differences in Health*, Nutrition and Population, HNP Poverty Thematic Group of the World Bank.

Hawley, W.A., P.A. Phillips-Howard, F.O. ter Kuile, D.J. Terlouw, J.M. Vulule, M. Ombok, B.L. Nahlen, J.E. Gimnig, S.K. Kariuki, M.S. Kolczak and A.W. Hightower (2003). 'Community-wide effects of permethrin-treated mosquito nets on child mortality and malaria morbidity in western Kenya'. *Am J Trop Med Hyg*, 68:121-127.

Howard, S.C., J. Omumbo, C. Nevill, E.S. Some, C.A. Donnelly and R.W. Snow (2000). 'Evidence for a mass community effect of insecticide-treated mosquito nets on the incidence of malaria on the Kenyan coast'. *Trans.R.Soc.Trop.Med.Hyg* 94: 357-360.

Indome, F., B. MacLeod, J.F. Phillips and F. Binka (1995). *The HRS Technical Manual. Version 2.0*, 1-51. New York, USA, The Population Council.

Koenig, M.A., D. Bishai and M.A. Khan (2000). *Child Survival Interventions and Health Equity: Evidence from Matlab, Bangladesh*. Unpublished Paper, Johns Hopkins University, Department of Population and Family Health Sciences.

Lengeler, C. (1998). *Insecticide-treated mosquito nets and curtains for malaria control – A Cochrane Review*. 1-54. Basel, Switzerland, Swiss Tropical Institute.

Mwageni, E., D. Momburi, Z. Juma, M. Irema, H. Masanja and the TEHIP and AMMP Teams (2002). Rufiji DSS. In: *INDEPTH Monograph Series: Population and Health in Developing Countries. Volume 1*. Mortality in INDEPTH DSS Sites. IDRC Press. Ottawa.

Phillips, J.F. and B.B. MacLeod (1995). 'The rapid transfer of demographic suveillance systems with automated software generation technology'. A final report to the Thrasher Research Fund.

Kakwani, N., A. Wagstaff and E. van Doorslaer (1997). 'Socioeconomic inequalities in health: measurement, computation and statistical inference'. *Journal of Econometrics* 77 87-103.

Rustein, S.O. (1984). 'Socio-economic differentials in infant and child mortality', *World Fertility Survey Comparative Studies No. 22* Voorburg, ISI.

Sahn, D.E. and D. Stifel (2000). *Assets as a Measure of Household Welfare in Developing Countries*, Unpublished Paper, Cornell University.

Wagstaff, A. and N. Watanbe (1999). *Inequalities in child malnutrition in the developing world*, DECRG, The World Bank, Washington, DC.

Woelk, G. and P. Chikuse (2000), *Using Demographic and Health Surveys (DHS) data to describe intra country inequities in health status: Zimbabwe*, Paper presented at the EQUINET Conference, Mid-Rand South Africa, 12-15th September 2000.

Annex 2.1 English Translation of Asset Survey Questionnaire Nested into a DSS Survey

Rufiji DSS Asset Survey, October 2000 – January 2001

Does anyone in this household own any of the following items?

1. Bicycle .... Y N
2. Car .... Y N
3. Motorbike .... Y N
4. Radio .... Y N
5. Refrigerator or freezer .... Y N
6. Television .... Y N
7. Clock/watch .... Y N
8. Own Land .... Y N
9. Sofa .... Y N
10. Wooden bed .... Y N
11. Electric Iron .... Y N
12. Mattress (foam/cotton) .... Y N
13. Own House .... Y N
14. Radio cassette .... Y N
15. Wardrobe .... Y N
16. Water pump .... Y N
17. Livestock .... Y N
18. Sewing machine .... Y N
19. Poultry .... Y N
20. Mosquito net .... Y N
21. Satellite dish .... Y N
22. Fan .... Y N
23. Do you rent this house .... Y N

What are the floors of this house made of? .... ☐
1=Earth, 2=Wood, 3=Tiles, 4=Cement, 5=Other

What are the walls of this house made of? .... ☐
1=Stone, Coral Block, Cement block, Burnt bricks, 2=Mud bricks (plastered or unplastered), wood, 3=Galvanized, mud & stick, mud, 4=Grass, Cardboard, 5=Other

What is the roof of this house made of? .... ☐
1=Tiles, concrete, cement, 2=Galvanized iron or asbestos,
3=Bamboo, wood, mud, grass, thatch, 4=Other

How many rooms are used for sleeping in this household? .... ☐

What is the main source of drinking water for this household? .... ☐
1=piped into residence, 2=rain water harvesting, 3=public tap, 4=vendor,
5=river, canal, spring, 6=other

What is the time in minutes to the main water source? .... ☐

What is the main toilet facility for this household? .... ☐
1=Private Flush, 2=Shared flush, 3=VIP or pit, 4=Neighbour or bush, 5=other

What is the main source of energy for cooking in this household? .... ☐
1=Electricity, propane, or solar, 2=biogas, kerosene or charcoal, 3=firewood,
4=crop residue, coconut husks, sawdust, animal dung, chaff, grass, 5=other

Annex 2.2   Distribution of assets and housing conditions by quintiles (%)

Quintiles (Per cent of Population)

Variable	Poorest 1st	2nd	3rd	4th	Least Poor 5th	Average	Poorest/Least Poor Ratio
If household has							
Bicycle	12.0	38.0	55.0	52.0	55.0	42.0	0.22
Car	0.0	0.0	0.0	0.0	2.0	0.0	0.00
Motorbike	0.0	0.0	0.0	0.0	1.0	0.0	0.00
Radio	10.0	50.0	64.0	61.0	76.0	52.0	0.13
Refrigerator	0.0	0.0	0.0	0.0	4.0	1.0	0.00
Television	0.0	0.0	0.0	0.0	3.0	1.0	0.00
Clock/watch	4.0	28.0	52.0	50.0	72.0	41.0	0.06
Land	99.0	98.0	93.0	89.0	73.0	91.0	1.36
Sofa set	0.0	0.0	0.0	2.0	23.0	5.0	0.00
Bed	99.0	97.0	97.0	96.0	97.0	97.0	1.02
Iron	0.0	0.0	5.0	14.0	37.0	11.0	0.00
Video	0.0	0.0	0.0	0.0	4.0	1.0	0.00
Matress	0.0	6.0	29.0	46.0	81.0	32.0	0.00
Wardrobe	0.0	0.0	0.0	5.0	33.0	8.0	0.00
Water pump	0.0	0.0	0.0	0.0	1.0	0.0	0.00
Livestock	0.0	0.0	1.0	2.0	8.0	2.0	0.00
Sewing machine	0.0	0.0	1.0	2.0	1.0	3.0	0.00
Poultry	57.0	59.0	55.0	46.0	36.0	51.0	1.58
Satellite dish	0.0	0.0	0.0	0.0	1.0	0.0	0.00
Fan	0.0	0.0	0.0	0.0	9.0	2.0	0.00
Owning house	99.0	94.0	88.0	80.0	55.0	83.0	1.80
Three sleeping rooms	17.0	24.0	26.0	24.0	21.0	22.0	0.81
Four sleeping rooms	6.0	13.0	15.0	21.0	15.0	14.0	0.40
Earth floor	100.0	100.0	100.0	93.0	28.0	84.0	3.57
Wooden floor	0.0	0.0	0.0	0.0	1.0	0.0	0.00
Tiled floor	0.0	0.0	0.0	1.0	1.0	0.0	0.00
Cement floor	0.0	0.0	0.0	2.0	54.0	11.0	0.00
Other floor	0.0	0.0	0.0	4.0	16.0	4.0	0.00
Stone walls	0.0	0.0	0.0	2.0	29.0	6.0	0.00
Bricks wall	0.0	0.0	1.0	3.0	7.0	2.0	0.00
Mud /stick wall	78.0	75.0	80.0	85.0	54.0	74.0	1.44
Grass wall	16.0	9.0	5.0	2.0	0.0	7.0	0.00
Other wall	5.0	15.0	14.0	8.0	10.0	11.0	0.50
Tiled roof	0.0	0.0	0.0	0.0	0.0	0.0	0.00
Asbestos	0.0	0.0	6.0	47.0	79.0	26.0	0.00
Thatch roof	100.0	99.0	89.0	45.0	17.0	70.0	5.88
Other roof	0.0	1.0	5.0	8.0	4.0	4.0	0.00
If household uses...							
Electric energy	0.0	0.0	0.0	1.0	1.0	0.0	0.00
Firewood	100.0	100.0	100.0	94.0	49.0	89.0	2.04
Kerosene	0.0	0.0	0.0	2.0	36.0	8.0	0.00
Residue energy	0.0	0.0	0.0	0.0	0.0	0.0	0.00
Other energy	0.0	0.0	0.0	3.0	14.0	3.0	0.00
If household water source is							
Piped into residence	0.0	0.0	0.0	0.0	1.0	0.0	0.00
Private well	0.0	0.0	0.0	1.0	1.0	0.0	0.00
Public well	100.0	99.0	98.0	97.0	88.0	97.0	1.14
Public tap	0.0	0.0	0.0	0.0	1.0	0.0	0.00
Vendor	0.0	0.0	0.0	1.0	8.0	2.0	0.00
River	0.0	0.0	1.0	1.0	1.0	1.0	0.00
If household's toilet is ...							
Flush toilet	0.0	0.0	0.0	0.0	0.0	0.0	0.00
VIP	0.0	0.0	0.0	0.0	1.0	0.0	0.00
Pit	80.0	95.0	95.0	96.0	97.0	92.0	0.82
Bush	14.0	3.0	3.0	2.0	1.0	4.0	14.00
Neighbour	6.0	2.0	2.0	2.0	2.0	3.0	3.00

Chapter 3

Child Health Inequity in Rural Tanzania: Can the National Millennium Development Goals Include the Poorest?

Rose Nathan, Joanna Armstrong-Schellenberg, Honorati Masanja, Sosthenes Charles, Oscar Mukasa and Hassan Mshinda

Summary

Health inequity can be assessed between various groups of people with different social attributes such as gender, ethnicity, income, geography, etc., through a methodology used to identify the poorest households and assessment of the inequity. People who belong on the lower side of those social attributes are considered as vulnerable. Here the poorest children living in the poorest households are focused on as a vulnerable group for which we provide empirical evidence of health inequity in comparison to those in the least poor households in rural Tanzania.

Both longitudinal and cross-sectional data were collected in the framework of the Ifakara Demographic Surveillance System. Household socio-economic surveys in 1997 and 2000 collected data on ownership of mosquito nets, radios, bicycles, house ownership status, tin roof, animals and poultry for nearly 12,000 households. Separate surveys asked questions on child health indicators for about 7,000 children born between January 1998 and August 2000. Mortality data was drawn from the DSS database for years 1999 and 2000.

We assessed socio-economic status by constructing a household 'wealth index' based on household asset ownership as proposed by Filmer and Pritchett (2001). This approach uses the Principal Component Analysis (PCA), which involves a mathematical procedure that transforms a number of (possibly) correlated variables into a (smaller) number of uncorrelated variables. Based on the constructed index, households were classified into wealth quintiles, rated as poorest, very poor, poor, less poor and least poor.

Equity was summarized by the ratio of the value of the indicator in the poorest to that in the least poor. Finally, we assessed the statistical significance of the relationship between the selected health indicators and socio-economic status using a logistic regression of each health indicator on socio-economic status quintile. Trends among the different socio-economic status quintiles were tested using a χ^2 test.

The constructed socio-economic status index showed differential levels of poverty within households in an area that is predominantly rural and covering an area of only

2440 Km². By splitting the index into quintiles, the mean socio-economic status score ranged from - 1.49 for the poorest households (first quintile) to 1.9 in the least poor households (fifth quintile).

Analysis of the socio-economic status index by sex of the head of household revealed a gender perspective of poverty. A significantly higher proportion of female-headed households belonged into the category of the poorest with a clear regressive line across the constructed wealth quintiles. In this regard, the approach of segmenting households by socio-economic status index qualifies the common argument that female-headed households are more vulnerable (economically as one aspect) than those headed by males.

Each child health indicator included in this study showed a gap in levels between the poorest and the least poor households. The disparity was highest and statistically significant for a child having an illiterate mother (poorest-least poor ratio of 1.8) followed by infant mortality rates with a ratio of 1.5. For net ownership analysis was done for two years, 1997 (baseline) and 2000 (3 years after). We observed a change of the poorest/least poor ratio from 0.4 in 1997 to 0.6 in 2000.

Background

Estimates of health in every country are generally based on averages across the population studied. Yet such estimates may obscure the health problems of the poor, particularly if these are different from those of the richer members of society (Gwatkin et al., 1999). The renewal of concern for poverty and equity has led to a number of recent studies on health inequality across and within developing countries and shown disparities in socio-economic status, ethnicity and income on several health outcome measures (eg. Brockerhoff and Hewett 2000, Wagstaff, 2000). The recent analysis of the Demographic and Health Surveys data from several developing countries has produced insightful results (www.worldbank.org/hnp).

Equity of health service delivery deserves particular priority in poor communities. When the World Bank introduced the concept of essential package of health services, efficiency and equity were identified as the most fundamental markers for the choice of interventions (World Bank, 1993). Public health care services are increasingly delivered through a cost sharing approach. Introduction of user financing in social service provision and health in particular has been reported to result in severely limited access particularly among the poor. For example Mwabu et al. (1995) reported a very large decline in health services utilisation in Kenya as a result of an absolute increase in fees. Similarly, increasing user fees resulted in a massive decline of the utilisation of health facilities in Zambia (Booth et al., 1995). It has been shown that the poor are more cost sensitive in their demand for medical care than the non-poor (Sauerborn et al., 1994).

In many developing countries assessing wealth is not simple, particularly where salaried employment is not common and people depend on subsistence agriculture for their food. Where income, consumption or expenditure data are not readily available, one possibility is to use a household asset index (e.g. Filmer and Pritchett, 2001). Although there are limitations associated with the measure of the low to high ratio, in particular its failure to take into account the health status of the middle quintiles, nevertheless it remains useful in measuring the relative gap in health between the poor and the least poor and it is readily interpretable (Sudhir et al., 2001).

Health indicators that are relevant for policy include 'process' indicators such as vaccination coverage or distance to health facilities, and 'outcome' indicators such as infant mortality or anthropometric status. Many developing countries do not have adequate vital registration systems and make more use of process indicators for analysis of health. However, process indicators do not necessarily reflect health outcomes. The INDEPTH network of field sites is increasingly able to provide reliable data on outcome indicators in these settings. Moreover, the network, through its field sites, has the potential for providing information on trends of equity over time and thus provides a useful platform for monitoring equity in health. The equity component in monitoring intervention programmes has rarely been considered.

Here we describe inequalities in health and wealth within a scattered community of subsistence farming families in rural Tanzania, explore the relationship between health and wealth indicators, and outline policy implications of our findings. Our analysis is based on existing data collected within the framework of the Ifakara Centre Demographic Surveillance System (IC-DSS).

Methods

Study Population

The Ifakara DSS covers a rural population of about 65,000 people living in 13,500 households in 25 villages of Kilombero and Ulanga districts in Morogoro region, southwestern Tanzania. The population is ethnically heterogeneous. Subsistence farming is the main activity followed by fishing and small-scale trading. The area has a network of health facilities, hospitals, health centres and dispensaries. Malaria is the leading cause of death in children under five. A detailed description of the study area is elsewhere (INDEPTH network 2002, pp. 159-154).

Data

Data used in this analysis are from various surveys and the core DSS database as explained below.

Household characteristics   The Ifakara DSS has three rounds in a year – i.e. every household is visited every four months. Once yearly (September-December), interviewers complete a household questionnaire that collects data about the socio-economic status of each household in the DSS area. In this analysis we used data from year 2000 for the construction of the wealth index. The 2000 survey covered 11,905 households. All indicators of the socio-economic status of the household were dichotomous (had at least one item): ownership of a bicycle (41%), a radio (38%), livestock (2%), poultry (51%), a tin roof (18%), and living in a rented house (6%). Additionally, occupation of the head of household was recorded and dichotomized.

Net ownership was analyzed for two years (1997 and 2000) because we wanted to examine the impact of social marketing programmes in addressing equity. For consistency we included only those households matched in the two socio-economic surveys and used the index of socio-economic status constructed for the year 2000.

Child health indicators   A cohort of about 7,000 children born between January 1998 and August 2000 is being followed up and information on their health status is collected every four months. Only those with complete information and whose households could be matched with the relevant socio-economic survey file were included in this analysis (for each indicator about 95% of the records matched). A question on whether the child slept under a net the night previous to the survey date gave us another indicator of preventive measure against child illness. We also included the literacy status of the child's carer (most often the mother) from the cohort.[1]

The core DSS database includes records of deaths of individuals in the surveillance system since September 1996. We included all under-5 deaths that occurred between 1997 and 2000. Infant and childhood mortality rates are used as mortality outcome indicators. A total of 1,077 deaths of children under five were observed. Infant deaths were 645 and the remaining 432 occurred to children aged between 1-4 years. The infant mortality was 100 per thousand live births and childhood mortality of 19.0 per thousand person years.

Location of a health facility   can be a powerful determinant of access to health services particularly in places where transport is difficult. The time to the nearest health facility is a result of both location and means of transport from home to the health facility. We used time to the nearest health facility as a proxy for accessibility to health services and child's vaccination status as a proxy for utilization.

Definition of Indicators

Infant mortality   ($_1q_0$) is the probability that a newly born child will die before reaching his/her first birthday. And childhood mortality, ($_4q_1$) is the probability that a child at exactly age one will die before reaching his/her fifth birthday.

Vaccination status   Proportion of children aged between 12 and 23 months fully vaccinated against each of the diseases. Here we report only BCG vaccine.

Time to the nearest health facility   Estimated time to reach the nearest heath facility reported verbally by the respondent.

---

1   The cohort study is ongoing and the data remain incomplete and **SHOULD NOT BE CITED WITHOUT PERMISSION.**

Poverty Measurement Methods

Economic status is measured by constructing a wealth index using asset ownership as proposed by Filmer and Pritchett (2001).[2] This approach uses the principal components analysis (PCA) where scoring factors of each asset are summed up to form the asset index. Principal component analysis uses a technique of extracting from a large number of variables those few linear combinations of variables that best capture common information.

The asset index for each household, $(H_i) = f_1*(x_{i1}-x_1)/(s_1) +...+ f_n*(x_{in}-x_n)/(s_n)$ Where f_1 is the scoring factor of the first asset, x_{i1} is the ith household value for the first asset and x_1 and s_1 are the mean and standard deviation of the first asset variable over all households.

We used the first principal component to construct our index. That component explained 15% of the total variability and the greatest weight was given to ownership of a radio (0.49), ownership of a bicycle (0.47) and a house with a tin roof (0.45). Subsistence farming and fishing as the occupation of the head of household had negative scoring factors. The first four eigen values were 1.9, 1.3, 1.1 and 1.0 and accounted for 15%, 10%, 8% and 9% of the variation respectively. Based on the constructed indices, the households were classified into wealth quintiles, which we rated as the poorest, very poor, poor, less poor and least poor. We used these relative terms as opposed to 'poor' and 'rich' because there are no households in this study, which could be classified as 'rich'. Table 3.1 shows ownership of selected assets in each wealth quintile. From this, a consistent correlation of the assets with the index is seen. For indicators that are child-specific the children were stratified into the quintiles based on the constructed indices (quintiles of individuals at risk).

Table 3.1 Selected variables by wealth quintiles

Socio-economic status quintile	Number of households N = 11905	Mean SES score	Household assets (%)			
			Radio	Tin roof	Bicycle	Employed
Poorest	2464 (20.7%)	-1.43	0.0	0.0	0.0	0.0
Very poor	2423 (20.3%)	-1.01	1.4	0.0	1.6	0.0
Poor	2261 (19.0%)	-0.26	35.7	8.2	47.2	0.0
Less poor	2532 (21.3%)	0.6	71.1	18.2	73.0	0.0
Least poor	2225 (18.7%)	2.2	84.8	69.1	85.3	15.1

---

2   Validation studies done in Nepal, Indonesia and Pakistan showed that the approach provided an index that was as good in predicting school enrolment differentials as the conventional approaches based on expenditure.

Analytical Methods

Cross tabulations and poorest/least poor ratios are used to assess the variations of the child health indicators within households in the quintiles of the constructed asset index. Equity was summarized by the ratio of the value of the indicator in the poorest to that in the least poor. Finally, we assessed statistical significance of the relationship between the selected health indicators and socio-economic status using logistic regression of each health indicator on socio-economic status quintile. And tested for linear trend using a χ^2 test for trend.

Mortality, as a final outcome is examined using survival techniques. Life table methods were used to convert the central mortality rates into probabilities of dying.

All the analysis was done in STATA 7.

Results

A total of 11,905 households from the year 2000 socio-economic survey were included in the analysis. However, the size was variable across the indicators as determined by source of data and completeness of the included variables. Female-headed households were 20.3% of all the households included in the analysis. While 25% of the poorest households were female-headed, only 9.7% of the least poor households had a female head.

All the child health indicators, time to the health facility, BCG vaccine, sleeping under a net, mother's literacy and infant mortality showed variations across the constructed household wealth index (Table 3.2). Although trends are not consistent in every indicator, each showed a clear and statistically significant inequity (except childhood mortality) between those children in the least poor and those in the poorest households, favouring those in the least poor households.

Overall coverage of BCG vaccine was 86%. Although results showed a statistically significant association between BCG vaccine and wealth status of the households ($LR\chi^2$ = 17.64, p<0.0001) the poorest – least poor ratio is close to unity (0.9) (Table 3.2).

Forty six percent of all children included in this analysis took more than one hour to reach the nearest health facility.[3] There was a statistically significant association between time to reach health facility and wealth status of the household. However, the poorest – least poor ratio is reasonably close to unity, at 0.9 (Table 3.2).

Almost twice as many children in the poorest households were cared for by illiterate mothers/carers than those in the least poor households (ratio: 1.8, $LR\chi^2$ = 9.1, p<0.05).

Sleeping under a mosquito net during the previous night was quite common (63%). However, less than one-third of the children in the poorest households reported of having slept under a net in the previous night, compared to almost three-quarters

---

3   Since time is self reported we are aware of the risk of introducing an error by using the cut off of one hour due to the tendency of rounding off to a complete hour. The data we had could not allow for an alternative.

Table 3.2   Child health indicators by wealth quintiles

Indicator	N	Percent of children by wealth quintiles						
		Poorest	Very poor	Poor	Less poor	Least poor	P-level	Poorest/least poor ratio
Accessibility and utilisation indicators								
Less than one hour from a health facility	6029	50	50	53	57	58	0.000	0.9
BCG vaccine	5390	83	84	86	86	89	0.001	0.9
Child slept under a net	6032	29	65	62	65	76	0.000	0.4
Maternal indicators								
Illiterate mother/carer	5987	20	18	18	15	11	0.000	1.8
Mortality indicators								
$_1q_0$	6410	109	104	104	78	74	0.05	1.5
se		(7)	(10)	(8)	(7)	(7)		
$_1q_4$	23163	89	83	92	96	69	0.84	1.3
se		(7)	(11)	(10)	(9)	(8)		

Chi-squared test for linear trend in proportions/rates

of those in the least poor households, resulting in a very low poorest/least poor ratio (0.4).

Table 3.2 shows fairly stable infant mortality rates for the first three groups of the households followed by a sharp drop in the less poor households. At infancy, a child in a household categorised as poorest had about 50% higher risk of dying compared to one in a least poor household (test for trend p = 0.05). Childhood mortality was fairly stable across the wealth quintiles except a noticeable drop in the least poor group. There was no significant association between socio-economic status and childhood (1-4) mortality.

In 1997, net ownership by households showed a clear gradual increase along all the quintiles of the wealth index, with only 325 (23 %) of the households in poorest household owning at least one net. A statistically significant association of net ownership and household wealth quintiles was observed (LRχ^2 = 1,011.9, p<0.0001), the ratio of the poorest – least poor was 0.3 (Table 3.3). After three years, the proportion of the poorest households with nets had increased from 23% to 53%. The proportion of the least poor households that owned at least one bed net increased

Table 3.3   Net ownership in households by wealth index – 1997 and 2000

Year	N	Percent of households owning at lest one net						Poorest/
		Poorest	Very poor	Poor	Less poor	Least poor	P-level	least poor ratio
1997	6919	23	26	36	44	60	0.000	0.4
2000		53	63	76	86	93	0.000	0.6

Chi-squared test for linear trend

to 93% over the same period of time. Although the association of net ownership and wealth quintiles in year 2000 was statistically significant (LRχ2 = 1,100.8, p<0.0001), the poorest-least poor ratio for ownership improved from 0.4 to 0.6 (Table 3.3), however the poorest-least poor difference in ratios was unchanged (Figure 3.2).

Discussion

We observed a statistically significant trend across the asset index quintiles for the infant mortality (p = 0.05). Infants in the poorest households experienced about 50% excess risk of death compared to those in the least poor households. For childhood mortality, we did not observe a statistical association with the household socio-economic status. Perhaps this can be explained by the fact that malaria is the major cause of infant deaths in the study site and other studies in the same area have shown that children in the poorest households were less likely to seek suitable treatment once ill (Schellenberg et al., 2003). Results from the analysis of the 1996 Demographic and Health Survey for rural Tanzania did not show a clear trend across the quintiles for infants nor for the children under five, nevertheless infants in the poorest households experienced nearly thirty percent excess mortality over their peers in the least poor households (Gwatkin et al., 2000).

Children in the poorest households were less likely to sleep under a net and were more likely to have an illiterate carer. Other child indicators (BCG and time to health facility) showed an association with socio-economic status but the poorest-least poor ratios were very close to one. The BCG vaccine results are consistent with those reported by the 1999 Tanzania Reproductive and Child survey showing very small variations of BCG coverage within various background characteristics (NBS 2000).

Policy Implications

When mortality reduction targets are set for countries, the reference is made on national averages without considering disparities within the population. Tanzania's target for instance is to reduce infant mortality by three quarters by year 2025 (http://www.tanzania.go.tz).

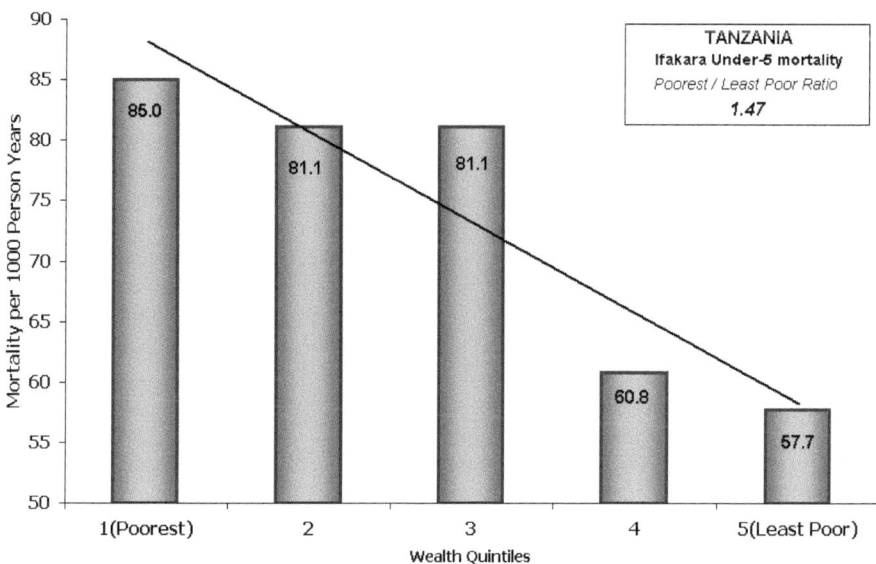

Figure 3.1   **Under-5 mortality by wealth quintile in the Ifakara DSS area 1997-2000**

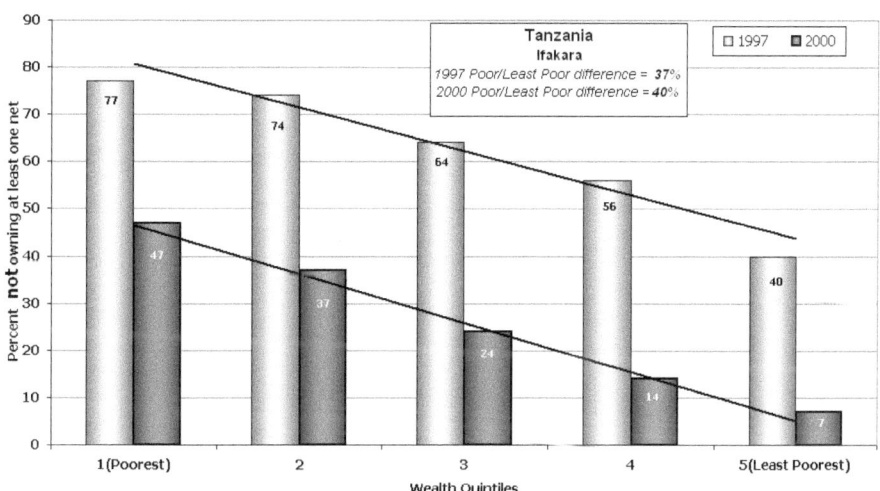

Figure 3.2   **Differences in household net coverage (expressed as non-coverage) by wealth quintiles in 1997 and 2000**

Based on the current level of 99 per 1,000 live births (NBS 2000) the target is to attain a level of 24.8 per 1,000 live births by year 2025. The poorest households in our study area are experiencing an infant mortality rate of 109 per 1,000 PYO and will therefore need to reduce the current mortality level rate by about 80% to achieve the national target by 2025. The challenge that faces implementers of national policies is how to target those groups (social or economic) which have a much longer way to go to attain the set levels. Furthermore, these targets might be achieved and celebrated without knowing whether the poorest groups have made any gain.

Poverty is more likely to be a major concern when considering ownership of nets rather than usage – once there is a net in the household there are many other factors that may determine who actually uses it. Among notable messages from this study is the improved accessibility to mosquito nets for all households and more importantly to the poorest households. The reduced gap between the poorest and the least poor is worthy of policy makers and health managers' attention. The nets were socially marketed from mid-1997 and sold using both public and private outlets. Perhaps this approach can be effective for reaching the poor with health services and disease control tools.

The approach used here to segment households by wealth status is simple and needs data that are easy to collect. By the use of this simple method we can contribute into the ongoing efforts to lessen human suffering from abject poverty, by bringing to light the existing inequity within small populations that seem homogenous. These results contribute to the most needed evidence base for efficient planning and advocacy. Since poverty is dynamic even in the absence of any deliberate intervention, monitoring it over time is necessary. Simple methods of this type can be used by planners and development programme managers to identify the poorest households, monitor and measure the impact of interventions in the specific constructed levels of poverty.

References

Booth D. et al., 'Coping with Cost Recovery', Report to SDA. Commissioned through the Development Studies Unit, Department of Social Anthropology. 1995, Stockholm University.

Brockerhoff M. and Hewett P. 'Inequality of child mortality among ethnic groups in sub-Saharan Africa'. Bulletin of the World Health Organisation 2000; 78 (1): 30-41.

Gwatkin D. R., Guillot M., Hueveline P. 'The burden of disease among the global poor'. Lancet 1999; 354: 586-89.

Gwatkin D.R., Rutstein S., Johnson K., Pande R.P. and Wagstaff A. *Socio-economic Differences in Health, Nutrition, and Population*. 2000, The World Bank.

INDEPTH Network. Ifakara DSS, Tanzania. In: 'Population and Health in Developing Countries. Population, Health, and Survival at INDEPTH Sites'. 2000; Vol. 1. 159-164. IDRC. Ottawa.

Mwabu G., Mwanzia J. and Liambila. 'User Charges in Government Health Facilities in Kenya: Effect of Attendance and Revenue', Health Policy & Planning 1995; 10(2): 164-70.

Filmer, D. and Pritchett, L. H. 'Estimating wealth effects without expenditure data – or tears: an application to educational enrolments in states of India'. *Demography* 2000; 38 115-32.

Sauerborn R., Nougtara A. and Latmer E. 'The Elasticity of Demand for Health Care in Burkina Faso: Differences across Age and Income Groups'. Health Policy & Planning 1994; 9(2):185-192.

Sudhir A., Diderichsen F., Evans T., Shkolnikov V.M. and Wirth M. In *Measuring Disparities in Health: Methods and Indicators. Challenging Inequalities in Health, From Ethics to Action.* Evans T., Whitehead M., Diderichsen F., Bhuiya A. and Wirth, Editors. 2001, Oxford University Press. pp. 48-67.

Armstrong Schellenberg J.A., Victora C.G., Mushi A., de Savigny D., Schellenberg D., Mshinda H. and Bryce J. 'Inequities among the very poor: health care for children in rural southern Tanzania'. Lancet 2003; 361: 561-66.

Tanzania Bureau of Statistics. *Reproductive and Child Health Survey – 1999.* 2000, National Bureau of Statistics and Macro International Inc.

World Bank. *Investing in health.* World Development Report. 1993, World Bank: Washington DC.

Wagstaff A. 'Child mortality across countries. Socio-economic inequalities in child mortality: Comparisons across nine developing countries'. Bulletin of the World Health Organisation, 2000. 78(1): 19-29.

Chapter 4

Health Inequalities
in the Kassena-Nankana District
of Northern Ghana

Cornelius Debpuur, Peter Wontuo, James Akazili and Philomena Nyarko

Summary

Improving health outcomes among the world's poor is currently a major priority in the international development community. In this regard, much of the focus has been on reducing deaths among young children. Although, there have been improvements in the health status of Ghanaians, the levels of infant and child mortality in Ghana remain unacceptably high. Besides, the national figures mask substantial differentials in health status among subgroups of the Ghanaian population. Child mortality for instance, ranges from 21 per 1,000 in the Greater Accra region to 109 per 1,000 in the Northern region. To facilitate policy and program efforts to further improve the health status of Ghanaians, it is important to identify those factors that produce differentials in health status.

This study explored the relationship between socio-economic status and health service utilization and health outcomes in the rural Kassena-Nankana district focusing on the most vulnerable members of the community – young children. The objective is to examine how gender, education and economic status relate to health care utilization and health outcomes in this rural community. We examined the relationship between socio-economic variables and health service utilization and health outcomes for children born between 1996-2000. Immunization status of children is used as a proxy for health service utilization and mortality under five years as an indicator of health outcome. Economic status was measured by a wealth index using compound assets and housing characteristics.

Results suggest that immunization status is better for children of educated mothers as well as those in the least poor compounds. Survival of children during the first five years is better for those who live within five kilometers of a health facility as well as those whose mothers have been to school. Sex of child has no effect on either immunization status or survival of children.

Background

Improving health outcomes among the world's poor is currently a major priority in the international development community. In this regard, much of the focus has been on reducing deaths among young children. Although, there have been improvements in the health status of Ghanaians, the levels of infant and child mortality in Ghana remain unacceptably high. For instance, infant mortality dropped from 133 per 1,000 in 1957 to about 77 per 1,000 in 1988 and then to 66 per 1,000 in 1998 (Ghana Statistical Service and Macro International Inc., 1999). However, these national figures mask substantial differentials in health status among subgroups of the Ghanaian population. Child mortality for instance, ranges from 21 per 1,000 in the Greater Accra region to 109 per 1,000 in the Northern region (Ghana Statistical Service and Macro International Inc., 1999). Economic disparities in survival are also glaring; the poor-rich ratio in under-5 mortality in Ghana is estimated to be 2.1 (Gwatkin et al., 2000). Such inequalities in health exist worldwide and reducing the poor-rich gap in health has become a priority to various organizations such as the World Health Organization and the World Bank as well as national governments.

Populations in rural communities are often assumed to be fairly homogenous. This perception coupled with the lack of relevant data have led to the absence of studies assessing socio-economic inequalities in rural Africa. Although socio-economic inequality may not be obvious in rural Africa, the capacity of individual households to deal with the challenges of life is not necessarily uniform. Variations in individual and household socio-economic conditions that shape behavior often result in variations in peoples' ability to protect and promote health. Consequently, variations in the health status of children may exist even in an apparently homogenous community. A recent study in rural Tanzania revealed poor-rich differences in care-seeking behavior in favor of the least poor (Schellenberg et al., 2003). To promote the health of rural communities therefore, it is essential that we identify subgroups that are disadvantaged so that health programs and interventions can be targeted at them, thereby making more focused use of available resources for health care. One way to identify disadvantaged groups within the community is to look at differentials in health status in relation to various socio-economic characteristics.

Inequalities in health reflect the influence of a number of socio-economic factors at the individual, household and community levels. However, much of the existing differentials in health status and health service utilization are often seen as unnecessary, avoidable and unfair. Indeed, the issue of health inequity largely derives from the belief that existing inequalities in health status are unnecessary and unfair. This underlies the growing concern over health inequity over the last decade.

In this chapter we explore the existence of health inequalities in a rural community in northern Ghana by examining the relationship between socio-economic characteristics and health care utilization and health outcomes in the Kassena-Nankana district. We focus on young children, as they are the most vulnerable members of the community. The objective is to examine how gender, education, and socio-economic status relate to health care utilization and health outcomes in a rural community in northern Ghana. Specifically, the study aims at the following objectives:

- Examine the relationship between various socio-economic status variables and health service utilization
- Examine the relationship between various socio-economic status variables and health outcome
- Identify the most and least disadvantaged groups in the Kassena-Nankana district in terms of service utilization and health outcomes.

The ultimate goal is to understand the role of individual and/or household factors in shaping health inequality in rural Ghana. So far studies on mortality differentials in Ghana have relied on national level data, and very few studies have analyzed mortality differentials within a single community. Thus this analysis is unique in its focus on a single district in Ghana, and the use of surveillance data. The overall objective of the Five Year Program of Work of the Ghana Ministry of Health is to reduce health inequalities in Ghana. This is part of the overall national poverty reduction strategy. A study such as ours could inform decisions and programs at the district level (where government policies eventually get translated into action) on how to overcome inequalities in health and improve the health status of rural residents.

The Study Setting

The setting for this study is the Kassena-Nankana district, located in the Upper East region of Ghana. This is mainly a rural community located in the north-eastern corner of the country bordering Burkina Faso and occupies an area of about 1675km^2. The area is relatively dry between October and April, with May to September as the rainy season. The predominant occupation is subsistence agriculture. As in many parts of rural Ghana, poverty is endemic in this district. Due to the erratic nature of rainfall and deteriorating soil quality, harvests are often poor and seasonal food shortages are not unusual.

Level of education in the district is low; about 64 percent of adults (15 years or older) in the district have had no formal education. In terms of gender more females (73%) than males (53%) have received no formal education (Nyarko et al., 2001). Both western and traditional practitioners provide health care. There is one district hospital, four health centers and several outreach stations. A recent attempt to restructure health delivery in the district has resulted in the relocation of some health staff to some of the villages.

The Navrongo Health Research Centre has been operating a demographic surveillance system for a population of about 140,000 in the Kassena-Nankana district since 1993. Mortality in the area is quite high; infant mortality in 1996 was estimated to be about 137 per 1,000 live births. Total fertility during the same year was around 4.5. However, both mortality and fertility have declined slightly. In 1999/2000, infant mortality rate was estimated to be 95.5 per 1,000 live births, while under-5 mortality stood at 164.5 per 1,000 (Nyarko et al., 2001).

Data and Analysis

The analysis is based on data from the Navrongo Demographic Surveillance System (NDSS). The NDSS is a longitudinal population registration system established in the Kassena-Nankana district in 1993 to support the research activities of the Navrongo Health Research Centre. Beginning with a baseline census in July 1993, demographic events have been continuously updated at 90-day intervals. Data are collected on births, deaths, in- and out-migration, marriages, and pregnancies. Basic demographic information such as sex, education, dates of events, and relationships of compound members is also included in the database. Details of the NDSS can be found elsewhere (see Nyarko et al., 2002; Binka et al., 1999).

In addition, socio-economic information on compounds is collected as part of the surveillance operations. Such information is only collected at the time that compounds are first registered in the NDSS. During Round 33 of the NDSS data collection (i.e. July-September 2001) information on compound ownership of various assets as well as characteristics of the compound was updated. Variables on which information was collected could be grouped into three. First are variables about compound ownership of various consumer durables (bicycle, radio, TV, sewing machine, etc). The second set of variables describes characteristics of the dwelling (toilet facility, source of drinking water, number of sleeping rooms, type of roof, etc.). Finally, the third set of variables captures ownership of domestic animals. For consumer durables and domestic animals the number of each item owned by compound members was recorded.

Since 1996, data on vaccination status of children under two years have been collected on an annual basis. The interviewer records dates of vaccination only if there is a health card or some other written record of the date of vaccination and the vaccination given. This means that some of the children classified as not vaccinated may in fact have received vaccinations but lost their health cards by the time of the survey. The vaccination data was last updated during Round 34 of NDSS data collection (i.e. October-December 2001).

For the analysis of health care utilization and health outcomes, we use children who were born in the district between 01 January 1996 and 31 December 2000 for whom we have information on vaccination, mother's identity and characteristics, as well as socio-economic data on their compounds. Children who were born elsewhere and migrated into the district have been excluded from the analysis.

We were able to identify a total of 19,640 children in the rural parts of the district who were born between 1996 and 2000. Of this number 10,711 (54.5%) were included in the vaccination analysis. 860 were excluded due to missing information on either the mother or compound of residence; 1427 were not found in the vaccination data base; 6089 could not present a health card; 216 had inconsistent information relating to date of birth and date when certain antigens were received; and 308 were excluded because they were below twelve months. For the mortality analysis, 18,751 (95.5%) of the identified births were included; 860 were excluded because of missing information on either the mother or compound of residence.

Variable Definitions

Health is an element of household welfare, and the health status of children reflects the welfare consequences of the household environment. Consequently, health care utilization in relation to children and the health status of children are the major outcomes of interest in this paper. Health behaviors such as immunization and modern curative care have demonstrable effects on child health and survival. The potential benefits range from protection against infection to recovery from infection. In view of the synergy between infection and nutritional intake, the prevention of infection is particularly important in promoting the health of children.

Immunization against the six childhood killer diseases is one of the effective ways of protecting children against infection and subsequently death. This makes vaccination coverage among children a useful indicator of health service utilization and one that is relevant for policy. We use vaccination status as a proxy for health service utilization. For the purposes of this analysis we used three measures to describe the vaccination status of children: whether or not a child has ever been vaccinated; whether a child received measles vaccination within the first year; and whether or not a child is fully vaccinated within the first year at the time of the last vaccination survey.[1] Children in Ghana are required to be fully vaccinated by the age of 12 months. 'Fully vaccinated' means that a child has received one dose each of BCG and measles, and three doses each of DPT and polio. The analysis of vaccination coverage is limited to children aged 12 months or above at the time of the last vaccination survey. In the NDSS, information on vaccination is collected only if a health card is seen. This means that there is no information on vaccination available for children who do not have vaccination cards or whose cards were not seen. Such children are excluded from the analysis. Annex 4.1 presents data on availability of health cards by background characteristic of children. Health cards were available for about 64% of the children in the vaccination data set.

Under-5 mortality is commonly used as an indicator in describing the health status of communities. Thus, we use survival during the first five years of life as our measure of the health status of children. This is considered as a much more objective measure than reported morbidity which is subject to reporting bias.

For the purposes of this study the key explanatory variables are: economic status of compound, education of mother, education of compound head, and sex of child. A wide range of factors affects health service utilization and health status. Consequently, we introduce other sets of variables as controls. These include year of birth of child, mother's age at birth of child, mother's marital status, ethnicity, compound size, distance of compound to a health facility, and health delivery strategy area.

The NHRC has geo-referenced all compounds and health facilities in the district; this makes it possible to calculate the distance of a compound to the nearest health facility using GIS. As part of the design of the Community Health and Family

---

1   According to WHO recommendations, children should be fully vaccinated by the end of the first year of life. Analysis of immunization coverage is therefore restricted to children older than eleven months.

Planning project (CHFP), the Kassena-Nankana district has been divided into four 'cells' with each cell representing alternative approaches to primary health care delivery in geographically contiguous areas. Each cell corresponds to the catchment area of a health center in the Kassena-Nankana district. In Cell 1, the major approach involves mobilizing traditional social organizational resources (known as 'Zurugelu') for health care delivery. The strategy for Cell 2 involves reorientation and relocation of community health nurses from the health center to live and work in the community under a new designation of community health officer. The approach in Cell 3 combines mobilization of social organizational resources with the reorientation and relocation of nurses to village locations. Cell 4 serves as the comparison area, where both the Zurugelu and relocated nurses are absent, and only the normal Ministry of Health approach to health delivery (i.e. a health center with outreach to communities) operates. Thus each compound in the district is located in an area that corresponds to a specific health delivery strategy.

Economic status of the compound is measured in terms of housing characteristics and assets per capita in a manner similar to the approach proposed by Filmer and Pritchett (1998, 1999) and used by others (e.g. Wagstaff and Watanbe, 1999; Rutstein, 1999). The approach involves the use of principal component analysis where the scoring factors of each asset are used to form an index for each household (A_j) based on the formula: $A_j = f_1*(a_{j1}-a_1)/(s_1) + ... + f_N*(a_{jN}-a_N)/(s_N)$ where f_1 is the 'scoring factor' for the first asset as determined by the procedure, a_{j1} is the jth household's value for the first asset, and a_1 and s_1 are the mean and standard deviation of the first asset variable over all households.

In the NDSS, data collection is done at the level of the residential unit or compound. Residential units (compounds) in the rural parts of the district typically consist of related individuals from several generations. Compounds vary in composition and size, and often comprise more than one nuclear family, ranging in size from 2 to 180 people, with an average of 12. In many cases, such residential units encompass more than one household (typically defined as a group of people who eat from a common pot). As a residential unit, the compound is not the equivalent of the household as is often used in socio-economic analysis. Compound members do not necessarily share housekeeping arrangements or eat from a common pot, and cannot be regarded as a single economic unit.

The method proposed by Filmer and Pritchett is based on asset data collected at the household and not at the dwelling level. However, the asset data used in this analysis were collected at the dwelling (compound) level and not the household level. For this reason we have expressed asset ownership on a per capita basis and used principal components to construct a wealth index in the manner proposed by Filmer and Pritchett (1988, 1999). The resulting wealth index of each compound was assigned to children below five years and the resulting population divided into quintiles ranging from the lowest twenty percent (Quintile 1) to the highest twenty percent (Quintile 5). The quintiles are used to represent a continuum of poverty from the poorest (Quintile 1) to the least poor (Quintile 5). A list of the assets and housing characteristics used in constructing the index and their factor scores are shown in Annex 4.2; the cut-off points for various quintiles are presented in Annex 4.3.

Methods of Analysis

Simple cross-tabulations and multivariate approaches are used to examine the links between socio-economic conditions and health outcomes. As an initial step we use cross-tabulations to explore the existence and magnitude of health disparities in the Kassena-Nankana district. Chi-squared tests are performed to assess the significance of these disparities. We use the poorest/least-poor ratio to indicate the wealth gaps in health care utilization and health outcomes. Multivariate analysis is used to further examine the association between our explanatory variables of interest and health outcomes after controlling for confounding factors.[2] All dependent variables are coded as dummy variables in the logistic regression; independent variables were entered into the equation as dummy variables, with the omitted category as the reference group.

Results

Health Service Utilization and Socio-economic Status

Of the 19,640 children born in the rural communities of the Kassena-Nankana district between 1996-2000 for whom we have relevant information, 10,711 (54.5%) were included in the analysis of vaccination coverage.[3] A distribution of the children included in the vaccination analysis by selected background characteristics appears in Table 4.1. Data on health service utilization (specifically immunization coverage) by various socio-economic characteristics are presented in Table 4.2. Overall, less than 10 percent of children twelve months or older have never been vaccinated. This may reflect the fact that only children whose health cards were available are included in the analysis. About 3 out of 4 children (75%) have received the measles vaccination; while slightly more than half (54%) have been completely vaccinated by age twelve months. The data suggest that children in the district are being vaccinated later than is recommended.

Socio-economic differences in vaccination coverage among children can be observed. In terms of wealth, immunization coverage appears to improve with wealth. Children in the poorest quintile are least likely to have received measles vaccination (with poorest/least poor ratio of 0.90), and are also less likely to have completed the required vaccinations by age twelve months compared to their counterparts in the least poor quintile (see Table 4.2).

Education appears to be related to children's immunization status. Children in compounds headed by someone who has been to school are more likely to be vaccinated than children whose compound heads have never attended school. Similarly, children with mothers who have ever attended school are also more likely to be vaccinated compared to those whose mothers have never been to school.

---

2   Results of the multivariate analysis are considered unsuitable for this volume and are thus not presented here.

3   Children below 12 months and those whose vaccination cards were not available have been excluded from the vaccination analysis.

Table 4.1   Distribution of children included in the vaccination analysis by selected socio-economic characteristics, Kassena-Nankana district[1]

Socio-economic characteristic	Number of births	Percent of children
Sex of child		
Female	5432	50.7
Male	5279	49.3
Child still alive		
No	436	4.1
Yes	10275	95.9
Child's Immunization status[2]		
Not immunized	730	6.8
Partially immunized	3864	36.1
Fully immunized	6117	57.1
Maternal age (years)		
Under 20	1148	10.7
20-29	4452	41.6
30+	5111	47.7
Mother's education		
No education	9085	84.8
Some education	1626	15.2
Mother's marital status		
Monogamous	5483	51.2
Polygamous	2303	21.5
Undefined	2925	27.3
Health delivery area		
Cell 1	1787	16.7
Cell 2	1672	15.6
Cell 3	3925	36.6
Cell 4	3327	31.1
Ethnicity		
Kassim	5537	51.7
Nankam/Other	5174	48.3
Compound size (# of people)		
Under 9	2491	23.3
9-13	2510	23.4
14-22	2965	27.7
Above 22	2745	25.6
Distance to nearest health facility		
Below 5 km	8357	78.0
5 or more km	2354	22.0
Education of compound head		
No education	9745	91.0
Some education	966	9.0

Table 4.1 continued

Socio-economic characteristic	Number of births	Percent of children
Wealth quintile of compound[3]		
Quintile 1 (Poorest)	2062	19.9
Quintile 2	2126	20.0
Quintile 3	2139	19.9
Quintile 4	2206	20.2
Quintile 5 (Least poor)	2175	20.0
Total	**10711**	**100.0**

Notes:
1 Table is based on children aged 12 months or older whose health cards were seen.
2 Based on vaccination status before the first birthday.
3 Three missing cases omitted.

Of the background characteristics examined in Table 4.2, only distance to a health facility and sex of child appear not to have a strong relationship with vaccination coverage. Since immunizations are not only given at the health facilities, the lack of relationship between distance to health facility and immunization coverage is not surprising. Indeed, the organization of outreach services for child immunization is aimed at minimizing the negative effect of living far away from the health facility. By taking immunization services outside the fixed health facility to outreach centers in the community, the distance required to get to the immunization point is drastically reduced. The lack of differential in vaccination coverage among male and female children suggests that there is no gender discrimination among children in terms of immunization in this society.

The relationship between socio-economic characteristics and immunization coverage is further examined using multivariate analysis. Table 4.3 presents unadjusted odds ratios for the three indicators of health service utilization for various quintile groups. Consistent with the pattern in Table 4.2 the odds of measles vaccination and complete vaccination before age 12 months significantly increase with wealth. For instance, a child in the least poor quintile is 63 percent more likely to have been vaccinated against measles compared to a child in the poorest quintile. The odds are even better for children in the least poor quintile when it comes to complete immunization before age twelve months.

When other factors are taken into account the effect of wealth on immunization status is reduced but not eliminated in the case of measles vaccination and complete immunization (results not shown). This suggests that the effects of other background characteristics do not eliminate the wealth effect on immunization coverage. Other factors that have significant effect on immunization coverage include mother's education, mother's marital status, education of compound head, compound size, and service delivery area (CHFP Cell) in which the compound is located. Children are more likely to be immunized when the mother has ever been to school, and also when the compound head has ever been to school. The location of the compound (in terms

Table 4.2 **Percentage of children 12 months and above who have never been vaccinated, percentage who received measles vaccination, and percentage fully vaccinated by age 12 months according to background characteristics**

Socio-economic characteristic	Percent never vaccinated	Percent vaccinated against measles by age 12 mths	Percent fully vaccinated by age 12 months	Number of children
Wealth Quintiles[1]	***	***	***	
Quintile 1 (Poorest)	7.4	73.5	48.8	2062
Quintile 2	7.1	75.8	55.4	2126
Quintile 3	7.9	76.6	55.8	2139
Quintile 4	5.7	79.3	59.4	2206
Quintile 5 (Least poor)	6.1	81.9	65.6	2175
Health delivery area	***	***	***	
Cell 1	5.4	77.7	62.4	1787
Cell 2	8.2	79.8	57.5	1672
Cell 3	5.7	82.8	63.6	3925
Cell 4	8.3	69.9	46.4	3327
Education of Cpd head	***	***	***	
None	7.0	77.2	56.4	9745
Some	4.9	80.4	64.0	966
Ethnicity	***	***	***	
Kassim	6.0	80.3	61.8	5537
Nankam/Others	7.6	74.5	52.1	5174
Compound size (# of people)	***	***	***	
Under 9	7.0	74.9	53.6	2491
9-13	8.4	76.0	55.3	2510
14-22	6.0	79.5	58.7	2965
Above 22	6.0	78.9	60.1	2745
Distance to health facility	ns	**	ns	
Below 5 km	6.6	77.0	57.3	8357
5 or more km	7.4	79.3	56.5	2354
Sex of child	ns	ns	ns	
Female	6.6	78.0	57.4	5432
Male	7.1	77.0	56.8	5279
Maternal age (years)	***	**	**	
Under 20	7.0	78.1	58.3	
20-29	6.0	79.4	60.2	1148
30+	7.5	75.6	54.2	4452
Mother's education	***	***	***	5111
None	7.3	76.4	55.1	9085
Some	4.2	83.3	68.2	1626

Table 4.2 continued

Socio-economic characteristic	Percent never vaccinated	Percent vaccinated against measles by age 12 mths	Percent fully vaccinated by age 12 months	Number of children
Mother's marital status	***	***		
Monogamous	6.5	78.2	57.8	5477
Polygamous	7.2	77.4	55.3	2307
Undetermined	7.1	76.2	57.3	2927
All children	**6.8**	**77.5**	**57.1**	**10711**

Notes:
ns = differences between groups not statistically significant at p = 0.05
*** p<0.001   ** p<0.01
1 Three missing cases omitted

Table 4.3   Unadjusted odds ratios of the likelihood of ever being vaccinated, receiving measles before age one, and complete vaccination before age one for asset index quintiles among children born between 1996-2000 in the Kassena-Nankana District

Wealth quintiles	Odds Ratio	z	p>z	95% CI
A. Odds ratios of ever been vaccinated				
Quintile 1 (Poorest)	1.00	–	–	–
Quintile 2	0.97	-0.18	0.857	0.74-1.29
Quintile 3	0.84	-1.30	0.193	0.65-1.09
Quintile 4	1.21	1.35	0.176	0.92-1.61
Quintile 5 (Least poor)	0.98	-0.14	0.889	0.75-1.28
B. Odds ratios of receiving measles vaccination before age 12 months				
Quintile 1 (Poorest)	1.00	–	–	–
Quintile 2	1.13	1.68	0.093	0.98-1.29
Quintile 3	1.18	2.32	0.020	1.03-1.36
Quintile 4	1.38	4.43	0.000	1.19-1.59
Quintile 5 (Least poor)	1.63	6.56	0.000	1.41-1.89
C. Odds ratios of full vaccination before age 12 months				
Quintile 1 (Poorest)	1.00	–	–	–
Quintile 2	1.30	4.22	0.000	1.15-1.48
Quintile 3	1.32	4.53	0.000	1.17-1.49
Quintile 4	1.53	6.93	0.000	1.36-1.73
Quintile 5 (Least poor)	1.99	10.98	0.000	1.77-2.26

Table 4.4   Background characteristics of children included in the mortality analysis, Kassena-Nankana District

Socio-economic characteristic	Number of births	Percent of births
Wealth quintile (4 missing cases omitted)		
Quintile 1 (Poorest)	3749	20.0
Quintile 2	3748	20.0
Quintile 3	3751	20.0
Quintile 4	3750	20.0
Quintile 5 (Least Poor)	3749	20.0
Health delivery area		
Cell 1	3031	16.2
Cell 2	2879	15.3
Cell 3	6533	34.8
Cell 4	6308	33.6
Education of compound head		
No education	17157	91.5
Some education	1594	8.5
Ethnicity		
Kassim	9388	50.1
Nankam/Other	9363	49.9
Compound size (# of people)		
Less than 9	4564	24.3
9-13	4417	23.6
14-22	5095	27.2
Over 22	4675	24.9
Distance to health facility		
Below 5 km	14487	77.3
5 or more km	4264	22.7
Child still alive?		
No	2588	13.8
Yes	16163	86.2
Sex of child		
Female	9484	50.6
Male	9267	49.4
Year of birth		
1996	3834	20.4
1997	3909	20.8
1998	3536	18.9
1999	3648	19.4
2000	3824	20.4
Maternal age (years)		
Under 20	2276	12.1
20-29	8024	42.8
30 +	8451	45.1

Table 4.4 continued

Socio-economic characteristic	Number of births	Percent of births
Education of mother		
No education	15955	85.1
Some education	2796	14.9
Mother's marital status		
Monogamous	9047	48.2
Polygamous	3964	21.1
Undefined*	5740	30.6
Total	**18751**	**100**

Note:
* Includes unmarried women as well as those whose marital status could not be determined due to insufficient information.

of service delivery area) significantly influences immunization status. Children located in Cell 4 (where routine health service activities are going on) are significantly less likely to be immunized than children in areas where additional interventions in health care delivery are taking place.

In summary, our results suggest a significant relationship between socio-economic status and immunization coverage. Household economic status, mother's education, and education of the compound head are the main socio-economic variables that have significant effects even when other factors are controlled. More importantly however, our results show that other factors may reduce but do not eliminate the effect of economic status on immunization.

Health Outcome (Under-5 Mortality) and Socio-economic Status

The survival status of children under five years is our main indicator of health outcome (Figure 4.1). Of interest here is whether or not the child survived the first five years and the extent to which variations in child survival relate to socio-economic status. Of the 19,640 children born between 1996 and 2000, 18,751 of them are included in the mortality analysis. Data on the background characteristics of these children are presented in Table 4.4. Of this number 2,588 (about 14 percent) were no longer alive as at December 2001. The children included in the mortality analysis were distributed among 8,087 compounds. Table 4.5 presents data on the distribution of deaths among children under five years according to selected background characteristics. In terms of economic status, there appears to be a steady decline in the proportion of children dead as wealth increases, with the poorest quintile recording 14.2 percent of deaths compared to 12.6 percent in the least poor quintile. With the exception of ethnicity and sex of child, significant variations in the proportions of children dead can be observed among the other variables in Table 4.5.

However, these seeming wealth differentials in mortality do not persist when the effects of other factors are taken into account. Although the hazard ratio decreases

Table 4.5   Percentage distribution of under-5 deaths among children born between 1996-2000 according to background characteristics

Socio-economic characteristic	Percent who have died	Number of children
Wealth Quintiles[1]	14.2	3749
Quintile 1 (Poorest)	14.6	3746
Quintile 2	14.2	3750
Quintile 3	13.2	3750
Quintile 4	12.6	3749
Quintile 5 (Least poor)		
Health delivery area	*	
Cell 1	14.4	3029
Cell 2	14.2	2878
Cell 3	12.7	6533
Cell 4	14.4	6308
Education of Compound Head	*	
No education	14.0	17155
Some Education	11.9	1593
Ethnicity	ns	
Kassim	14.1	9385
Nankam/Others	13.5	9363
Compound size (# of people)	***	
Under 9	15.9	4563
9-13	13.8	4417
14-22	12.9	5094
Above 22	12.6	4674
Distance to nearest health facility	*	
Less than 5 km	13.5	14484
5 or more km	14.9	4264
Sex of child	ns	
Female	13.6	9484
Male	14.0	9264
Year of birth	***	
1996	17.8	3831
1997	15.6	3909
1998	14.9	3536
1999	12.4	3648
2000	8.2	3824
Maternal age	***	
Under 20 years	16.9	2276
20-29 years	12.7	8021
30 and above	14.0	8451
Mother's Education	**	
No education	14.1	15952
Some Education	11.9	2796

Table 4.5 continued

Socio-economic characteristic	Percent who have died	Number of children
Mother's union status	***	
Monogamous	12.9	9049
Polygamous	13.7	3962
Undefined	15.2	5737
All children	**13.8**	**18748**

Notes:
ns = differences between groups not significant at p<0.05
*** p<0.001   ** p<0.01   * p<0.05
1 Four missing cases omitted.

with economic status, the differences are not significant (results not presented). Among the variables included in the analysis, mother's education, maternal age, ethnicity, distance to health facility and compound size appear to have a significant effect on mortality during the first five years. Thus it appears that other factors are much more important in explaining under-5 mortality differentials in the Kassena-Nankana district than economic status *per se*.

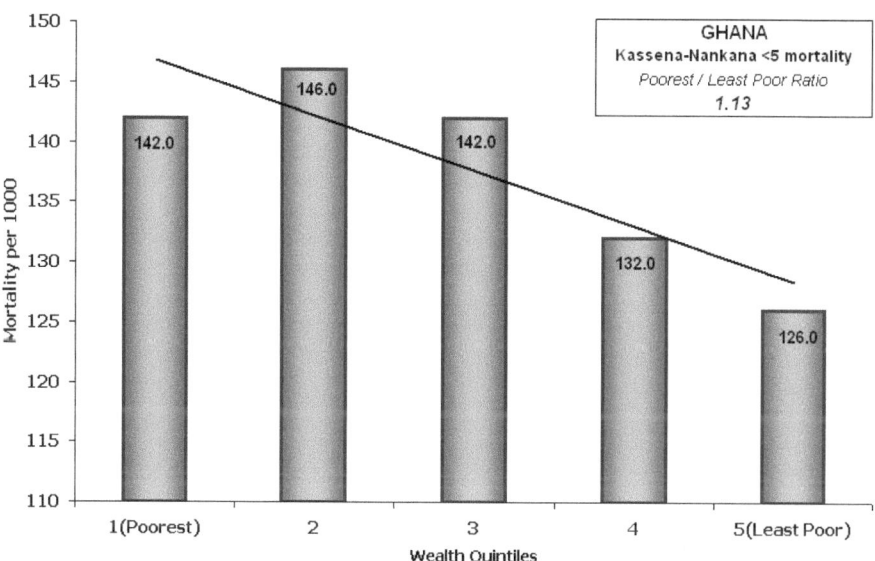

Figure 4.1   Under-5 mortality in Kassena-Nankana by wealth quintile

Discussion and Conclusions

We explored the relationship between socio-economic status and health service utilization and health outcomes in the rural Kassena-Nankana district, focusing on the most vulnerable members of the community – young children. Our objective was to examine how gender, education and economic status relate to health care utilization and health outcomes in this rural community. We examined the relationship between socio-economic variables and health service utilization and health outcomes for children born between 1996-2000 using immunization status of children and under-5 mortality as indicators of health service utilization and health outcome respectively. Economic status was measured by a wealth index combining assets per capita and housing characteristics in a manner akin to that proposed by Filmer and Pritchett (1998). The key findings from our analysis may be summarized as follows:

- Better immunization for children of educated mothers
- Better immunization for children in relatively wealthier compounds
- Better survival for children of educated mothers
- No significant differentials in survival of children in terms of economic status
- Sex of child has no relationship to either health service utilization or health outcome

From our results it appears that children in the poorest wealth quintile, children whose mothers have never attended school as well as those living in compounds headed by someone who has never been to school are the most vulnerable in terms of health service utilization. In terms of mortality however, children of teenage mothers, unmarried women, and children living more than five kilometers away from a health facility are the most vulnerable.

Using immunization status of children as an indicator of health service utilization, better immunization of children of educated mothers suggests that health service utilization is higher among the educated. This may reflect their understanding and appreciation of the importance of preventive measures in promoting the health of their children. In this case, health education among those without formal education would need to be intensified. However, given the poor timing of immunization, efforts should be made to extend the coverage of EPI activities. This not only entails making sure that children are immunized, but getting them immunized at the appropriate ages. Here also, health education at antenatal clinics, EPI centres and health durbars will be essential. Community involvement and support in legitimizing immunization campaigns and other outreach programs will go some way to ensure that children are immunized timely.

The wealth differential in immunization in favor of the relatively wealthy in this rural community is worrisome. In Ghana, childhood immunizations are free, and one would have thought that this would make it easily accessible to all in the community. However, as the data suggest, not everyone has equal access and utilization is obviously not equal. Exactly what drives these inequalities needs to be explored. One possible explanation may be found in the health seeking attitudes and practices of poor households. It may also be the case that certain indirect costs (in terms of travel to the immunization centre, time lost for other activities, etc.) incurred by households make it difficult for some households to avail themselves of services that exist in the community. Better immunization coverage among relatively better off households would suggest that such households are better able to absorb the indirect costs of getting children immunized. In this regard efforts to minimize the indirect cost of immunizations (especially to poorer households) may go some way to reduce the inequalities in service utilization. An increase in the number of immunization centres would greatly reduce travel cost as well as the time spent at these centres.

The survival advantage associated with mother's education calls for expansion in female education. However, this is a long-term approach that will benefit children of future mothers. In the immediate run it will be necessary to step up health education and health outreach activities in the community. The target of such activities will have to be extended to include not only mothers but also men and the community at large. Although mothers are responsible for childcare, their relationship with significant others define the limits and possibilities of healthy behavior. Thus health activities that do not involve these significant others may not achieve the desired results. Health programs that strengthen the capacity of mothers by providing them and their families with information, skills, resources and technologies to promote child health will need to be implemented.

Above all, the results of the analysis do suggest the existence of inequalities in health utilization and health outcomes in a generally poor society such as the Kassena-Nankana district. Thus even though people in the district are generally poor some are doing better than others when it comes to health care utilization and health outcomes, especially for children. In this regard, steps need to be taken to expand the availability of health and social services. In so doing attention should be paid to sections of the population that experience poorer health outcomes.

This study was meant as an exploratory exercise in measuring economic status and examining its effects on health service utilization and health outcomes in the rural Kassena-Nankana district. Measurement of economic status in poverty endemic communities is of interest to researchers and policy makers alike. Thus, this paper represents the beginning of a more systematic investigation of economic status and its effects on various outcomes in rural communities in Ghana. Following this exploratory exercise there is the need to refine our measurement of economic status. Especially in the Navrongo setting, this would require collection of household level data in a manner similar to the Demographic and Health Surveys. The availability of such data should make it possible to extend the investigation of inequity beyond health to include other outcomes such as education (given the significant relationship between education and several welfare indicators).

To strengthen our understanding of economic status in rural Ghana, the socio-cultural perceptions of wealth in specific communities need to be examined. Given that the current approach to measuring economic status is largely dependent on the use of household possessions it will be relevant to understand the importance communities attach to the possession of various items. This will be useful in determining which items to include in constructing a wealth index and in modifying our current data collection instruments and hopefully, put us in a better position to contribute towards monitoring the impact of national policies in reducing inequalities.

Acknowledgements

The research for this chapter was conducted as part of the INDEPTH Health Equity research program. We are grateful to the INDEPTH Network for the financial and technical support. We also wish to thank the NDSS team of the Navrongo Health Research Centre for the effort in collecting and processing the data used in this study. The Rockefeller Foundation's support in providing funding for the Navrongo DSS is gratefully acknowledged. Finally, we wish thank the reviewers of the initial report submitted to INDEPTH.

References

Binka F.N., P. Ngom, J.F. Phillips, K.F. Adazu and B. Macleod (1999). 'Assessing population dynamics in a rural African society: Findings from the Navrongo demographic surveillance system'. *Journal of Biosocial Science*, 31:375-391.

Filmer, D and L. Pritchett (1998). 'Estimating Wealth Effects without Expenditure data – or Tears: An Application to Educational Enrollments in the States of India'. World Bank Policy Research Working Paper No. 1994.

Ghana Statistical Service (GSS) and Macro International Inc. (MI) (1999). *Ghana Demographic and Health Survey 1998*. Calverton, Maryland: GSS and MI.

Gwatkin, D.R., S. Rutstein, K. Johnson, P.R. Pande, and A. Wagstaff (2000). *Socio-economic Differences in Health, Nutrition, and Population in Ghana*. HNP/Poverty Thematic Group of The World Bank. May 2000.

Gwatkin D.R. (2000). 'Health inequalities and the health of the poor: What do you know? What can we do?' Bulletin of the WHO; 78(1):3-18.

Nyarko, P., P. Wontuo, G. Wak, D. Chirawurah, P. Welaga, S. Nchor, A. Hodgson and F. Binka (2001). The Navrongo Demographic Surveillance System 2001 Report to the Rockefeller Foundation. Community Health and Family Planning Project Documentation Note Number 45.

Nyarko, P., P. Wontuo, A. Nazzar, J. Phillips, P. Ngom and F. Binka (2002). Navrongo DSS, Ghana. In *Population and health in developing countries. Volume 1. Population, health, and survival at INDEPTH sites*. INDEPTH Network 2002.

Schellenberg, J.A. et al. (2003). 'Inequalities among the very poor: health care for children in rural southern Tanzania' The Lancet. Published online February 4, 2003 http://image.thelancet.com/extras/02art2280web.pdf

Wagstaff, A. (2000). 'Socioeconomic inequalities in child mortality: Comparisons across nine developing countries'. Bulletin of the WHO; 78(1)19-29.

Annex 4.1   Availability of health card among children born between 1996-2000 by background characteristics, Kassena-Nankana District

Socio-economic characteristic	Card not seen (%)	Card seen (%)	Number of children
Wealth Quintiles[1]			
Quintile 1 (Poorest)	37.8	62.2	3317
Quintile 2	36.1	63.9	3329
Quintile 3	35.4	64.6	3309
Quintile 4	34.2	65.8	3353
Quintile 5 (Least poor)	34.7	65.3	3329
Health delivery area			
Cell 1	33.0	67.0	2667
Cell 2	34.3	65.7	2546
Cell 3	32.5	67.5	5818
Cell 4	40.7	59.3	5610
Education of Cpd head			
None	36.0	64.0	15230
Some	31.5	68.5	1411
Ethnicity			
Kassim	33.4	66.6	8309
Nankam/Others	37.9	62.1	8332
# of people in compound			
Under 9	37.9	62.1	4009
9-13	35.6	64.4	3900
14-22	34.8	65.2	4546
Above 22	34.4	65.6	4186
Distance to health facility			
Within 5 km	35.2	64.8	12902
Above 5 km	37.0	63.0	3739
Sex of child			
Female	35.6	64.4	8433
Male	35.7	64.3	8208
Maternal age			
Under 20	40.5	59.4	1931
20-29	37.3	62.7	7105
30+	32.8	67.2	7605
Mother's education			
No education	36.0	64.0	14192
Some education	33.6	66.4	2449
Mother's marital status			
Monogamous	32.4	67.6	8105
Polygamous	35.4	64.6	3571
Undefined	41.0	58.9	4965
All children	**35.6**	**64.4**	**16641**

Note:
1 Four missing cases omitted.

Annex 4.2   List of assets and factor scores

Asset variable	Factor score
Has modern design	0.08
Has zinc roof	0.07
Source of water – pipe-borne	0.06
Source of water – borehole water	−0.01
Number sleeping rooms	0.10
Number of bicycles	0.11
Number of beds	0.28
Number of donkey carts	0.05
Number of radios	0.28
Number of sewing machines	0.10
Number of lamps	0.13
Number of coal pots	0.17
Number of cows	0.04
Number of sheep	0.06
Number of goats	0.07
Number of pigs	0.05

Annex 4.3   Cut-off points for wealth quintiles

Wealth quintile	Asset index value	
	Lowest	**Highest**
Quintile 1 (Poorest)	Lowest	−2.04665
Quintile 2	−2.04665	−1.18968
Qunitile 3	−1.18968	−0.24373
Quintile 4	−0.24373	1.34299
Quintile 5 (Least poor)	1.34299	Highest

Socio-economic Status and Child Mortality in a Rural Sub-District of South Africa

Kathleen Kahn, Mark Collinson, James Hargreaves, Sam Clark and Stephen Tollman

Summary

This study examines equity issues related to infant and child mortality rates during 1992-2000 in a rural border-region of South Africa, an area characterised by significant poverty differentials and migration. Child mortality rates among different socio-economic strata, and between the South African and Mozambican sub-populations have been compared. Various possible risk factors for child mortality are analysed and discussed. The study included an open cohort of all children who spent time in the Agincourt sub-district during the period 01/01/1992 to 31/10/2000, an area comprising 21 villages and 11,500 households, with a population of some 70,000 people situated in the Bohlabela district of South Africa's rural north-east, adjacent to the country's border with Mozambique.

The Agincourt Health and Demographic Surveillance System provides nine years of population surveillance data together with a household-level index of economic status with which to test hypotheses concerning the relationship between socio-economic status and childhood mortality. An index of household economic status was constructed by combining the variables from the household asset survey (conducted in 2001), and a principal component factor analysis was performed to determine the relevant weights to assign to each variable. The study cohort included 30,733 children in 10,353 households.

The relationships between gender and nationality of household head, our chosen poverty markers, and the economic index indicate that both did reflect economic status, but that the nationality of household head was a better discriminator of poverty. The mothers in female-headed households tended to give birth at more vulnerable ages (12-19 and 40-49), tended to be better-educated, were more prone to migrancy and were much less likely to be Mozambican. High-risk factors emerging from these rural South African analyses indicate the need for broad socio-economic development as well as specific public health interventions. Of particular importance in the Agincourt analyses is an excess mortality among children born into a particularly marginalized group, namely Mozambican households located in the former 'refugee' settlements of South Africa's Limpopo Province. This result calls for further investigation into the factors that mediate this difference.

Background

Inequalities of personal health status may be biologically or genetically determined and consequently inevitable. Health inequities, in contrast, result from differences in health status outcomes between groups that are avoidable and unnecessary, and hence unacceptable and unjust (Whitehead, 1992). Widening health gaps among groups or nations call for global, national and sub-national responses that involve the public health and development communities. They also call for careful monitoring and evaluation of interventions, not just for their impact on health outcomes but also their impact with respect to these 'equity gaps'. While disproportionate health improvements for the most disadvantaged are often necessary to decrease inequity between groups, more advantaged groups tend to benefit first (Lanata, 2001; Victora et al, 2001). Consequently, successful public health interventions can contribute perversely to widening of the equity gap, at least in the shorter term. Research describing and measuring differential health status between groups is vital if these are to be redressed through health and development policies.

Following decades of government legitimised discrimination against black communities and households during the 'apartheid' era, South Africa remains with a legacy of marked income, living standards and health disparities. The South African government has committed itself to poverty alleviation, with closing the health, education and income equity gaps a primary goal of its health and social policy. However the impact of the government's current macroeconomic framework – the Growth, Employment and Redistribution strategy (GEAR) – on economic and social change is unknown (Gilson and McIntrytre, 2001). In a 1998 report to the South African government, the most significant indicators of poverty in South Africa were found to be race, gender and educational level of household head, lack of employment, and rural and Provincial residence (Gilson and McIntrytre, 2001). Black African[1] rural households in the poorer provinces, with a female head without secondary school education, and high unemployment, are likely to be particularly vulnerable.

The Agincourt sub-district of South Africa's largely rural Limpopo Province was part of a former Bantustan 'homeland' that today scores high on all these poverty indicators. South Africa's political and economic history, particularly through the first half of the 20th century, saw forced resettlement and concentration of black Africans into geographically dispersed and ecomically inhospitable 'homelands'. These areas were turned into cash economies by restricting access to land, conducting a process of 'villagisation', and imposing taxes on the rural population. The aim was to provide labour for the country's mining and industrial sectors, but the outcomes

---

1   The term 'African' indicates the terminology used as part of Apartheid legislation. Its use here does not imply acceptance of the former racist stratification of the South African population.

were entrenched migration, spousal separation and spatially divided or 'stretched' households, a pattern that remains common today. Remittances back to local households tend to create inequalities in these labour migrant sending communities (Massey, 1987; Curran, 2001). The Agincourt sub-district, together with other areas along the north-eastern border of South Africa, is also characterised by the presence of Mozambican nationals, many of whom fled their country's civil war in the early to mid-1980s and who have remained in South Africa as so-called 'self-settled' refugees. While local integration has been identified as a durable solution in certain refugee situations (Kibreab 1989), the Mozambicans have experienced variable legal, economic and social incorporation in South Africa, and hence remain a vulnerable subgroup.

This study examines infant and child mortality rates during 1992-2000 in a rural border-region of South Africa. The region is characterised by significant poverty differentials and the presence of Mozambican residents. Child mortality rates between different socio-economic strata, and between the South African and Mozambican sub-populations, are compared. Various possible risk factors for child mortality are analysed and discussed.

Methods

Study Area

The Agincourt sub-district, comprising 21 villages and 11,500 households, with a population of some 70,000 people, is situated in the Bohlabela district of South Africa's rural north-east, adjacent to the country's border with Mozambique (Figure 5.1). Nearly a third of the population are of Mozambican origin. The dependency ratio is high, with 44% of the population under 15 years of age and 4% over 65 years.

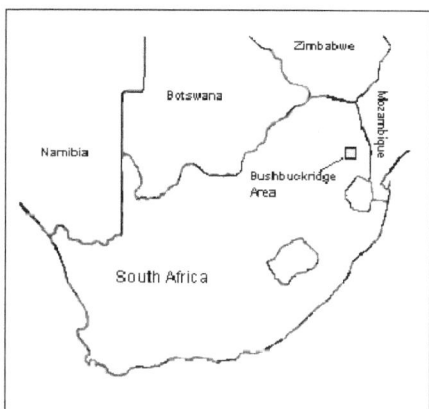
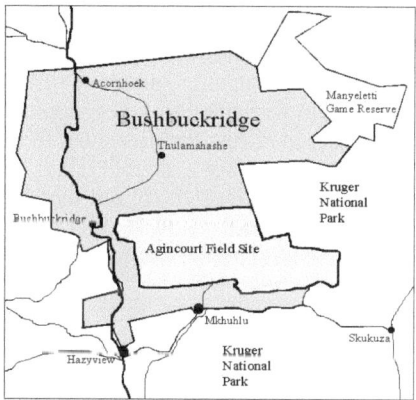

Figure 5.1   Map indicating location of Agincourt subdistrict, Bushbuckridge region, Limpopo Province, South Africa

Labour migration is extensive as local employment opportunities are few. The area, which is densely populated (148 persons per square kilometre) and arid with low rainfall, does not support subsistence agriculture. Around 60% of men and 14% of women 30-49 years migrate from the area to work elsewhere for at least 6 months of every year. Although school enrolment often occurs late, 85% of children aged 10-14 years enter primary school. Less than 50% continue to secondary school, however, and only 3% receive post-secondary education (Tollman et al., 1999). A health centre and five satellite clinics are situated within the study villages; three district hospitals are located 25 to 60 kms away. In general, people tend to seek care from a combination of allopathic and traditional health care providers.

Agincourt Health and Demographic Surveillance System

The Agincourt Health and Demographic Surveillance System (HDSS) monitors key demographic events and socio-economic variables in the Agincourt sub-district, an area lacking vital registration. Details of the HDSS study design and methods have been published previously (Collinson et al., 2002). A baseline census was conducted in 1992, followed by annual census update rounds with comprehensive recording of vital events. Variables measured routinely by the HDSS include: births, deaths, in- and out-migrations, household relationships, resident status, refugee status, education, antenatal and delivery health-seeking practices (Tollman, 1999; Tollman et al., 1999; Collinson et al., 2001). The data model for household membership accounts for the high levels of circular migration in this population by including on the household roster non-resident members who retain significant contact with the rural home (Collinson et al., 2001). The household head is the 'main household decision-maker' as reported by the household respondent. Given this definition, it is possible for the household head to be absent from the household for a majority of the time.

During the census update rounds a trained, lay fieldworker interviews the most knowledgeable respondent available at the time of visit to each household. Individual-level information is checked for every household member. All events that have occurred since the previous census update are recorded. Where appropriate, certain questions are directed at specific household members, for example maternity history or pregnancy outcome information is asked directly from the woman involved; a verbal autopsy is conducted with the person most closely involved with the deceased during their terminal illness. Revisits are undertaken when appropriate respondents are not available. Data quality checks include duplicate surveying of a random sample of 2% of households. In addition, a number of validation checks are built into both fieldwork and the data-entry process. The software system used consists of a relational database constructed in Microsoft SQL Server.

Study Cohort

The study included an open cohort of all children who spent time in the Agincourt sub-district during the period 01/01/1992 to 31/10/2000. Entry into the cohort was captured by entry to the HDSS (through birth, in-migration or presence in the baseline census), and exit from the cohort was determined either by exit from the HDSS before 5 years of age (through death or out-migration), turning five years during the follow-up period, or reaching the end of the follow-up period before the fifth birthday. 30,733 children were included in the cohort analysis contributing a total of 81,017 person years of follow-up.

Poverty Measures

During the 2001 census update, a household asset survey was conducted. This recorded salient features of the living conditions and assets of each household in the surveillance population. The questionnaire contained 34 ordinal variables, covering such areas as building materials and structure of the main dwelling, access to water and electricity, and ownership of appliances, transport and livestock. Variables were developed through a process of discussion and refinement with local field staff and community members. Several iterations of questionnaire piloting were conducted in the study site and elsewhere in the district.

An index of household economic status was constructed by combining the variables from the household asset survey and conducting a principal component factor analysis to determine the relevant weights to assign to each variable. The model that used the first principal component was selected because it summarised the most information across the variables and, in subsequent testing, correlated best with the individual variables making up the score. The first principal component was divided into quintiles labelled: low, medium-low, medium, medium-high and high economic status.

A similar household economic status index has been reported to discriminate effectively among levels of economic status in other rural African settings (Schellenberg et al., 2003). A limitation in our study, however, was that the variable was only available for 2001. It was therefore an inadequate measure for a longitudinal design going back several years, due to possible fluctuations in household economic status over the observation period. We used the economic indicator, therefore, to evaluate suitable proxy measures that could represent economic status in a longitudinal design: gender (May et al., 1995) and nationality (Dolan et al., 1997) of household head.

Household headship tends to be strongly gendered, with a male generally made head if one is in the household. Adult males are more frequently breadwinners than females, and when adult females are employed they often earn lower wages. A male-headed household, therefore, has the opportunity for a higher 'male' income, while a female-headed household is more likely to have a lower income and at least one less income-earning adult.

Nationality of household head relates to poverty through a combination of social discrimination, limited legal status, and restricted access to services including housing, water, education, pensions, and health services. Thus the indicators chosen to reflect poverty related to (1) lack of assimilation and integration in society (of Mozambicans), and (2) gender discrimination (of females in a household and in the labour force).

Analytic Methods

A longitudinal 'survival' analysis was undertaken to examine the effects of economic status variables on infant and child mortality over the observation period. Mortality rates were calculated as deaths per person years of follow-up, and were assessed for different socio-economic groups. Several known risk factors for childhood mortality were also investigated. These included maternal factors (age, education, marital status and presence in the home during the child's early years), household factors (sex of household head, presence of other female adults in the household, and number of co-resident children) and health service utilisation factors (mother receiving antenatal care and childbirth in a health facility). Most risk factors were measured at the time of cohort entry. Exceptions were the age of the child's mother, measured at birth, the marital status and presence of the mother after the child was born (measured at the census after cohort entry), and the educational status of the mother (measured in 1992 and 1997, or at the time of in-migration). Data on health service utilisation was only available for those children born into the cohort.

Poisson regression was used for risk factor analysis with the relative risks described as incident rate ratios. These are described for mortality over the first year of life, for ages 1-5 years, and for the full follow-up period (Annex 5.1). Stata 7.0 was used for analysis (Stata Corporation 2001).

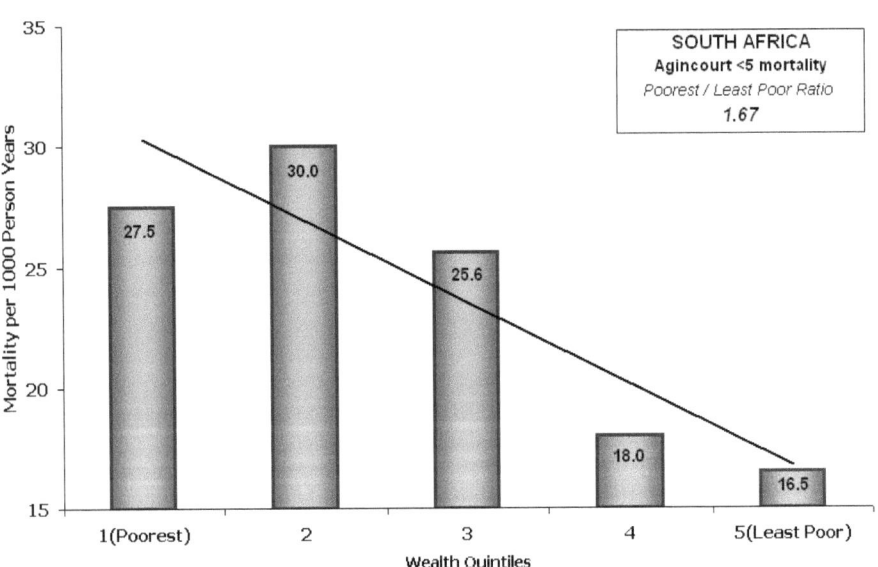

Figure 5.2   Age specific death rates of children under 5 by wealth quintile, Agincourt, 2000-2001

Results

Economic Status and Child Mortality

Although not available for the study cohort period, the socio economic index does correlate with under five mortality in 2000-2001 (Figure 5.2), with the highest economic class showing the lowest under five mortality (17/1000) and the two lowest classes showing the highest under five mortality (28/1000 and 30/1000 respectively).

The study cohort included 30,733 children in 10,353 households. Of the total number of households, 70.7% were male-headed and 29.3% female-headed. South African-headed households comprised 66% with 32% headed by a Mozambican. Missing values for nationality of household head accounted for the remaining 2%.

Figure 5.3 demonstrates the extent to which the chosen poverty markers, gender and nationality of household head, correlate with the economic index. The figure presents the proportion of female-headed and Mozambican-headed households in each economic stratum as determined by the asset index, with a vertical line indicating the confidence interval. Nationality of head was a good discriminator of economic status, with the highest proportion of Mozambican-headed households in

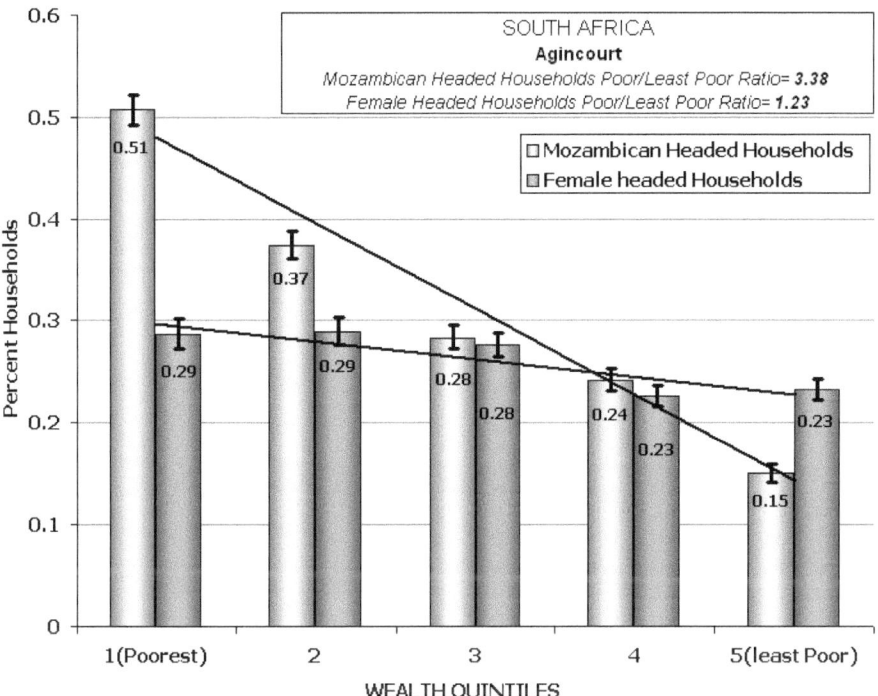

Figure 5.3   Proportion Mozambican- and female-headed households by wealth quintile, Agincourt 2001. 95% confidence intervals shown

the lowest economic class. This proportion dropped roughly linearly as the economic situation improved. The confidence intervals do not overlap, indicating that the proportion of Mozambican-headed households in each stratum is significantly different. The relationship between the proportion of female-headed households and the economic strata is less strong, although there is a significantly higher proportion of female-headed households in the lowest three economic strata compared with the two highest categories.

Figure 5.4 shows that both nationality of household head and gender of household head were factors associated with a differential in childhood mortality. The differential was more pronounced in the case of Mozambican compared with South African headed households, with the former showing a childhood mortality rate of 8.6 deaths per 1000 child years at risk, and the latter 6.2 deaths per 1000 child years at risk. The difference in mortality rate between female and male-headed households was evident, though not as marked. The childhood mortality rate in female-headed household was 8.0 deaths per 1000 child years at risk compared with 6.8 deaths per 1000 child years at risk in male-headed households. (Confidence intervals of incidence rate ratios are presented in Annex 5.1. The confidence interval for sex of household head is approaching significance, while that for nationality of household head is statistically significant.)

Children from Mozambican households are seen to have a worse mortality profile than their South African counterparts, although the risk of mortality associated with living in a Mozambican household was not equally distributed across all ages. Children aged 0-1 years had the same mortality rate in both Mozambican and South African households, while the mortality rate improved more rapidly in subsequent age groups in South African compared to Mozambican households, with a significantly

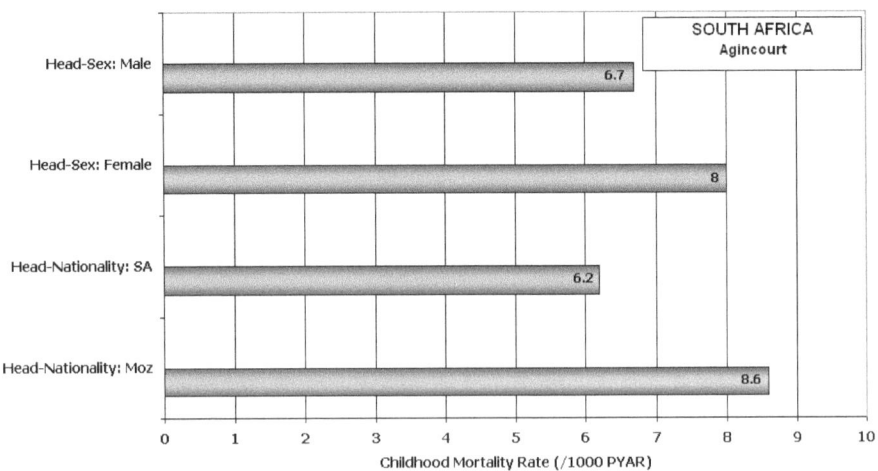

Figure 5.4   Childhood (<5yrs) mortality rates in the categories of poverty indicators: sex and nationality of household head. Agincourt 1992-2000

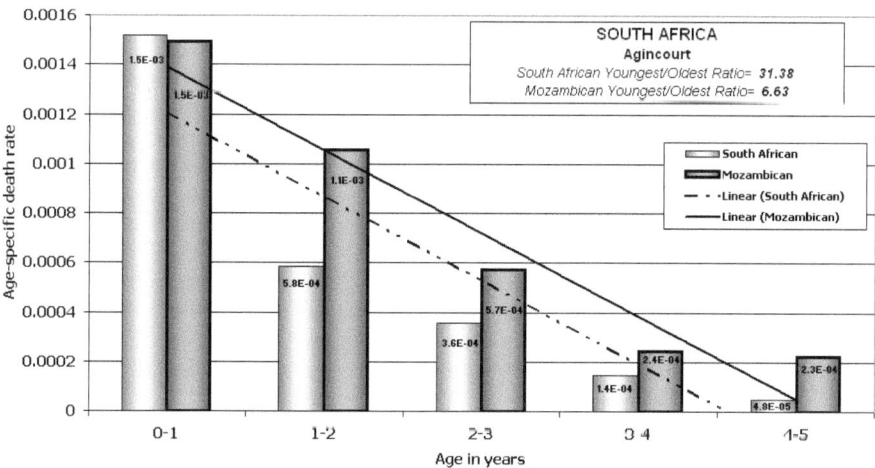

Figure 5.5   Age-specific death rates by nationality of household head, Agincourt, 1992-2000

higher mortality in Mozambican compared to South African children from age 1 onwards. Children aged 1-5 years in Mozambican headed households have almost a 90% greater chance of dying than children from South African headed households. (Figure 5.5; confidence intervals in Annex 5.1).

Annex 5.2 provides information on differences in maternal characteristics between male and female-headed households, and between South African and Mozambican headed households. Female-headed households differed significantly in composition from male-headed households in ways that could hypothetically impact on childhood mortality. Female-headed households had significantly higher proportions of mothers who were very young (adolescent) or old (over the age of 40 years), who had secondary or tertiary education, who were single (never married or marriage dissolved), and who migrated for work. They were far less likely to be Mozambican. Multivariate analyses (details available from authors) examining the influence of gender of household head together with mother's age and marital status, showed that gender of head was relatively unimportant compared to the impacts of maternal age and marital status. The child of an unmarried mother was found to be at highest risk of dying by their sixth birthday.

Mozambican headed households had a significantly higher proportion of mothers with no education and who had not given birth within a health care facility. In contrast, South African headed households had significantly higher proportions of mothers who were adolescent or over the age of 40 years, who never married, and who migrated for work.

Risk Factors for Childhood Mortality

Annex 5.1 provides details of the risk factors for childhood mortality under investigation in the longitudinal survival analysis. The socio-economic variables examined were associated to differing extents with childhood mortality.

The sex of the child did not appear to significantly interact with the chance of survival. Children born to mothers aged 20-29 years, the commonest childbearing age, seemed to have a survival advantage over children born to both older and younger mothers in the first year. However, children born to the oldest mothers who survived this first year exhibited the lowest mortality rates during the next 4 years. While there was surprisingly little variation in mortality rates between children born to mothers with either no schooling or some formal schooling, it appears that children born to the small group of mothers who go on to post-school education had much improved survival chances.

Children born to unmarried mothers were at greater risk. Additionally, the death of a mother during the period between cohort entry and the next census was strongly predictive of the child's mortality status. The effect was particularly strong during the infancy period, but was also seen during the period from the 1st to the 5th birthday. Part of the excess mortality is likely due to neglect following death of a mother, and part to child morbidity – mothers who died of HIV/AIDS may have had HIV infected babies. Children whose mothers were migrant during the year after their birth generally enjoyed a slightly improved survival profile. Children from large families, both those with a large number of other adult females, and those with many co-resident children, were at greater mortality risk, particularly after infancy.

The overwhelming majority of mothers had some contact with an antenatal clinic during their pregnancy, but children born to the small group without contact tended to be at increased risk of dying (although this effect did not quite reach statistical significance). There was little evidence to suggest that being born in a health facility had a significant impact on the child's survival chances after birth.

Discussion

Much of the literature on health equity describes disparities between nations although more recent work is emerging on equity gaps within countries. Inequality in under-5 mortality was found across nine developing countries, with inequalities highest in Brazil, high in Nicaragua and the Philippines, intermediate in Cote d'Ivoire, Nepal and South Africa, and lowest in Ghana, Pakistan and Vietnam (Wagstaff, 2001). A within-country study, examining child mortality disparities between ethnic groups in eleven countries of sub-Saharan Africa, found ethnic mortality differences closely linked to economic inequity in many of the countries, and to differential use of health services in the Sahel region. These findings indicate that analyses of child mortality in Africa should be conducted within a social and epidemiological framework. (Brockerhoff and Hewett, 2001). Far less effort to-date, however, has been devoted to the measurement of health equity at the more microeconomic level of the household. This study contributes to a broader effort to fill this information gap.

The Agincourt Health and Demographic Surveillance System provides nine years of population surveillance data together with a household-level index of economic status with which to test hypotheses concerning the relationship between socio-economic status and childhood mortality. There are, however, limits to the study imposed by the data and the analytic methods used. The estimates of mortality are lower than estimates for other HDSS sites (INDEPTH 2001). While this may reflect the comparatively favourable conditions in Agincourt compared to other rural African sites, and despite concerted efforts to promote full reporting, the possibility of underreporting of postnatal deaths cannot be ruled out. Certain exposure variables had to be fixed at time of child's entry into the cohort or the time of the following census because data on these variables were not available over the observation period. While we built each variable to provide the most accurate measure of risk over the key period of vulnerability, hence capturing the most important value, smaller changes in risk over time may not be adequately represented. Another limitation arises from the subset of cohort children to whom we could not match a mother (just under 6% of the study population). These cases are ignored in statistical analyses involving mothers' characteristics that may introduce bias in the estimations.

The relationships between gender and nationality of household head, our chosen poverty markers, and the economic index indicate that both did reflect economic status (Figure 5.3), but that the nationality of household head was a better discriminator of poverty.

Female-headed Households

Using gender of household head as a proxy for socio-economic status involves an assumption that female-headed households are systematically of a lower socio-economic status. While this suggestion is backed by analyses conducted on South African data (May, 1995; Lestrade-Jefferis, 1996), other work suggests that this may not always be the case (Garey and Townsend, 1996; Peters, 1983; Posel, 2001). In this study, female-headed households differed significantly in composition from male-headed households (Annex 5.2). The mothers in female-headed households tended to give birth at more vulnerable ages (12-19 and 40-49), tended to be better-educated, more prone to migrancy and were much less likely to be Mozambican. Probably the most striking difference between female and male-headed households was the marital status of mothers. Gender of head, it seems, could equally be a proxy for mothers being unmarried and/or mothers being very young or very old. Multivariate analyses indicate that age and marital status were more important in explaining variance in childhood mortality than simply the gender of household head in the household of the child's birth. Mother's marital status emerged from these analyses as a key social factor conferring protection on children, suggesting that formalising a union in this society goes a long way towards social protection and support.

Multiple causal mechanisms, therefore, appear to mediate the effect of gender of household head on child mortality. Very young mothers are biologically vulnerable and less mature caregivers; unmarried mothers suffer the social stigma of a child born out of wedlock with ensuing reduction in social support; while female-headed households are on average straitened by at least one adult male salary, as well as having less access to resources and less power to influence their situation in general.

Mozambican-headed Households

Children born into Mozambican households exhibited a significantly worse mortality profile than their South African counterparts. While Mozambican mothers were almost as likely to visit an antenatal clinic during their pregnancy, their children were much less likely to be born in a health facility. Additionally, the mother's of Mozambican children were much less likely to have received any formal education (Annex 5.2). These findings suggest that access to public sector services for this group may be limited by legal and logistical factors. Both poor health care utilisation and lower educational levels might be expected to increase mortality among Mozambican children. However, the evidence for such an effect is not strong. Multivariate analyses produced little evidence that differentials in childhood mortality between different nationality groups are the result of differences in any of the risk factors investigated in this study. Rather, if anything, some protective factors are concentrated among the Mozambican group, including significantly fewer adolescent and unmarried mothers. Despite this, the effect of Mozambican nationality had a powerful influence on childhood mortality.

On differentiating the relationship between an effect on infant mortality and one on mortality in the ages 1-5 years, it appears that the excess mortality of Mozambican children was concentrated in the later age period. There is some evidence to suggest that while almost all children are breastfed, Mozambican children may be breastfed for longer, especially given the much lower proportion of Mozambican mothers who are part of the migrant labour force. This may have provided additional protection for Mozambican infants, who may otherwise have exhibited high mortality in this age group as well.

Mozambicans are a marginalised group in this society, having entered South Africa as refugees. Qualitative work points to the fact that living conditions in predominantly Mozambican villages are far worse than those in other villages; the refugee settlements generally have no schools, the quality of housing is considerably worse and they are particularly isolated from public transport (Hargreaves, 2000). An environmental survey conducted in 1993 demonstrated that exclusively refugee villages were consistently worse off than 'mixed' villages with respect to access to water and fuel, and the types of sanitation and waste disposal methods used (Dolan et al., 1997). The authors point to the lack of legal status as a major contributor to the Mozambican's vulnerability. Exploitation in the workplace reflects their economic vulnerability demonstrated by lower consumption of cash goods, with free firewood used in favour of more convenient forms of energy. Legal vulnerability and consequent lack of certainty over the future results in less investment in domestic infrastructure such as pit latrines, as well as lack of government and NGO investment in public infrastructure such as water supply and electricity. Incorporation of the refugee villages into health and social services development is clearly indicated from both a human rights and public health perspective (Dolan et al., 1997).

Residents in the refugee villages are also socially isolated. For many, their relative social isolation compounds past experience of conflict, war, flight, and family disruption. The psychosocial impact of these experiences on individuals and families is compounded by break-up of traditional communities and erosion of social networks. Social marginalisation, together with life experience and character and personality of the mother, can lead to diminished maternal self-esteem and maternal depression. These can in turn result in child neglect and abuse (Lanata, 2001).

Implications of Study Findings

Gaps persist in our understanding of what factors contribute to the excess mortality among children from different backgrounds. Cause of death data may be useful in unravelling the relationships. More research could be conducted to further quantitatively investigate the pathways through which children are at greater risk (quality of housing, parental neglect, abuse). The relationship between maternal self-esteem and maternal depression with child health and development requires investigation. Combined study designs, whereby qualitative methods provide deeper insight into quantitative findings, and quantitative methods improve the generalisability of qualitative results (Lancet, 1994), could yield greater understanding of the pathways involved and lead to intervention strategies that target high risk populations, or aim to reduce the level of key risk factors among the general population.

However an equity-oriented approach may not give priority to such research. While much recent scientific enquiry has taken a 'high risk' perspective, there have been calls for a refocusing of public health science. Rose (1985) has pointed out that public health programmes may be better aimed at the general population rather than specific risk groups. In the same way, it has been suggested that by targeting distal factors on a causal pathway, small changes in 'relative risk' may result in large benefits for population health. While there is a temptation to view the results of population differentials in health status as primarily 'hypothesis generating', there is a need to re-evaluate the implications of these studies for health and social interventions. This is not to suggest that more complete information would not be useful. However, the constant search for more proximate causes of health differentials may obscure the search for public health solutions that directly target distal causes of health inequality.

It is necessary to proceed cautiously. Public health and development interventions that target the general population may produce positive health outcomes, but may also *increase* equity gaps, reaching the better-off first and taking time to percolate through to the more disadvantaged (Lanata, 2001; Victora, 2001). Despite this, improved child survival, with concurrent narrowing of socio-economic and gender disparities, has been reported in Bangladesh (Bhuiya et al., 2001). Here greatest improvements in child mortality were seen in the lowest socio-economic strata, with poor girls experiencing the greatest health gains. Bhuiya et al. (2001) point out the difficulty in determining if the health gains were due to particular public health interventions (such as immunisation or the promotion of oral rehydration therapy), or represent the outcome of secular changes in poverty, behaviour, maternal education and environmental factors. It seems clear, however, that health and socio-economic development together confer the greatest survival advantage on children.

High-risk factors emerging from these rural South African analyses indicate the need for broad socio-economic development as well as specific public health interventions. Improvements in family planning services can limit the number of very young mothers and provide greater choice to young adult women. While labour migration of mothers appeared to confer some protection, presumably through regular remittances, employment opportunities closer to their rural homes would enable women to breastfeed for longer. South Africa has a far-reaching system of social grants targeting children, the elderly and those with serious physical and mental disabilities. Many families eligible for social support grants do not receive them and systems promoting their uptake are urgently needed. Better access to transport, education and health services are called for, particularly in the more underserved villages.

Of particular importance in the Agincourt analyses is an excess mortality among children born into a particularly marginalized group, namely Mozambican households located in the former 'refugee' settlements of South Africa's Limpopo Province. This excess mortality is not adequately explained by differential socio-economic risk profiles as measured in this study, and hence may not be addressed by the interventions suggested above. This result calls for further investigation into the factors that mediate this difference. On a broader level, the finding calls for interventions that facilitate integration of former Mozambican refugees into South African society. Two recent amendments to government policy are now contributing to this: the opportunity for former Mozambican refugees to apply for permanent residence in South Africa, and efforts for them to gain access to a range of social support grants. While such interventions are beyond the realm of standard public health strategies, they are precisely the kinds of intervention that may have greatest impact when population health equity is the prime outcome of importance.

Acknowledgements

Core support for the Agincourt Health and Demographic Surveillance System is through a grant from the Wellcome Trust, number 069683/Z/02/Z. The Andrew Mellon Foundation supports migration research in the Agincourt Unit. All activities of the Unit depend on support of the study communities. This work was conducted with funding from the Rockefeller Foundation and World Bank through the INDEPTH Network. Ethical clearance was granted by the University of the Witwatersrand's Committee for Research on Human Subjects [Medical] (No. M960720).

References

Bhuiya, A., Chowdhury, M., Ahmed, F. and Adams, A.M. (2001). 'Bangladesh: An intervention study of factors underlying increasing equity in child survival'. In: Evans, T., Whitehead, M., Diderichsen, F., Bhuiya, A. and Wirth, M. (eds). *Challenging inequities in health: From ethics to action.* New York: Oxford University Press.

Brockerhoff, M. and Hewett, P. (2000). 'Inequality of child mortality in sub-Saharan Africa'. *Bulletin of the World Health Organisation* 78(1):30-41.

Collinson, M.A., Tollman, S.M., Garenne, M. and Kahn, K. (2001). 'Temporary female migration and labour force participation in rural South Africa'. A working paper of the Agincourt Health and Population Unit, University of the Witwatersrand and Princeton University's Center for Migration and Development.

Collinson, M.A., Mokoena, O., Mgiba, N., Kahn, K., Tollman, S.M., Garenne, M., Herbst, K., Malomane, E. and Shackleton, S. (2002). 'Agincourt DSS, South Africa'. In: Sankoh et al.

(eds). *Population and Health in Developing Countries. Volume 1. Population, Health and Survival at INDEPTH Sites.* Ottawa: IDRC. Chapter 16.

Curran, S.R. and Saguy, A.C. (2001). 'Migration and cultural change: A role for gender and social networks?' *Journal of International Woman's Studies* 2(3).

Dolan, C., Tollman, S., Nkuna, V. and Gear, J. (1997). 'The links between legal status and environmental health: A case study of Mozambican refugees and their hosts in the Mpumalanga (Eastern Transvaal) Lowveld, South Africa'. *International Journal of Health and Human Rights* 2(2):62-84.

Garey, A. and Townsend, N. (1996). 'Kinship, courtship, and child maintenance law in Botswana'. *Journal of Family and Economic Issues* 17(2):189-302.

Gilson, L. and McIntyre, D. (2001). 'South Africa: Addressing the Legacy of Apartheid'. In: Evans, T., Whitehead, M., Diderichsen, F., Bhuiya, A. and Wirth, M. (eds). *Challenging inequities in health: From ethics to action.* New York: Oxford University Press.

Hargreaves, J. (2000). 'Village Typology in the Agincourt sub-district'. Unpublished report. Agincourt Health and Population Unit, School of Public Health, University of the Witwatersrand, South Africa.

INDEPTH (2002). 'Comparing Mortality Patterns in *INDEPTH* Sites'. In: Sankoh et al. (eds). *Populations and Health in Developing Countries. Volume 1. Population, Health and Survival at INDEPTH Sites.* Ottawa: IDRC. Chapter 6.

Kahn, K., Tollman, S.M., Garenne, M., Gear, J.S.S. (1999). 'Who dies from what? Determining cause of death in South Africa's rural northeast'. *Tropical Medicine and International Health* 4:433-441.

Kibreab, G. (1989). 'Local settlements in Africa: a misconceived option?' *J Ref Stud* 2:468-90.

Lestrade-Jefferis, J. (2000). 'The labour market'. In: Udjo, E. O. (ed). *The People of South Africa Population Census 1996.* Pretoria: Statistics South Africa.

Lanata, C.F. (2000). 'Children's health in developing countries: issues of coping, child neglect and marginalisation'. In: Leon, D. and Walt, G. (eds). *Poverty, Inequality and Health: An international perspective.* Oxford University Press.

Lancet (1994). 'Population health looking upstream'. Editorial. *Lancet* 343:429-430.

Massey, D. (1990). 'Social structure, household strategies, and the cumulative causation on migration'. *Population Index* 56(1): 3-26.

May, J., Carter, M. and Posel, D. (1995). 'The composition and persistence of poverty in rural South Africa: An entitlements approach'. Johannesburg: Land and Agriculture Policy Centre.

Peters, P. (1983). 'Gender, Developmental Cycles and Historical Process: A critique of recent research on women in Botswana'. *Journal of Southern African Studies* 10(1).

Posel, D.R. (2001). 'Who are the heads of household, what do they do, and is the concept of headship useful? An analysis of headship in South Africa'. Working paper. Department of Economics, University of Natal, Durban, South Africa.

Rose, G. (1988). 'Sick individuals and sick populations'. In: Buck, C., Llopis, A., Najera, E. and Terris, M. (eds). *The challenge of epidemiology.* Washington: Pan American Health Organisation.

Schellenberg, J.A., Victoria, C.G., Mushi, A., de Savigny, D., Schellenberg, D., Mshinda, H. and Bryce, J. (2003). 'Inequities among the very poor: health care for children in rural southern Tanzania'. *The Lancet* 361:561-566.

Stata Corporation (2001). Stata, Statistics – data analysis. Version 7.0. 4905 Lakeway Drive, College Station, Texas 77845, USA.

Tollman, S.M. (1999). 'The Agincourt Field Site – evolution and current status'. *South African Medical Journal* 89(8):855- 857.

Tollman, S.M., Herbst, K., Garenne, M., Gear, J.S.S. and Kahn, K. (1999). 'The Agincourt Demographic and Health Study – site description, baseline findings and implications'. *South African Medical Journal* 89(8):858-864.

Victora, C.G., Barros, F.C. and Vaughan, J.P. (2001). 'The impact of health interventions on inequalities: infant and child health in Brazil'. In: Leon and Walt, G. (eds). *Poverty, Inequality and Health: An international perspective*. Oxford University Press.

Wagstaff, A. (2000). 'Socioeconomic inequalities in child mortality'. *Bulletin of the World Health Organisation* 78(1):19-29.

Whitehead, M. (1992). 'The concepts and principles of equity and health'. *International Journal of Health Services* 22(3):429-445.

Annex 5.1   Univariate analysis of potential risk factors for childhood mortality

Factor		Number of subjects	Person Years at Risk (PYAR)	Rate (/1000 PYAR)	IRR (95% CI) 0-5 yrs	IRR (95% CI) 0-1 yrs	IRR (95% CI) 1-5 yrs
Sex of HH head	Male	22821	61071	6.8	1	1	1
	Female	7904	19922	8.0	1.18 (0.97-1.42)	1.25 (0.97-1.63)	1.09 (0.83-1.43)
Nationality of HH head	South African	18944	50125	6.2	1	1	1
	Mozambican	11418	30430	8.6	1.41 (1.19-1.68)	0.98 (0.78-1.26)	1.89 (1.49-2.39)
Sex of child	Male	15394	40645	7.3	1	1	1
	Female	15339	40371	6.9	0.95 (0.80-1.12)	0.92 (0.72-1.17)	0.97 (0.77-1.22)
Age of mother	12-19	5870	16628	8.1	1.20 (0.96-1.49)	1.25 (0.91-1.73)	1.09 (0.81-1.47)
	20-29	12719	34743	6.8	1	1	1
	30-39	6595	19472	6.5	0.96 (0.77-1.20)	1.00 (0.72-1.40)	0.93 (0.69-1.24)
	40 and over	5549	10174	7.8	1.16 (0.89-1.50)	1.68 (1.21-2.32)	0.58 (0.37-0.92)
Educational status of mother	None	8253	22593	6.5	1	1	1
	Primary	7403	20532	6.9	1.06 (0.84-1.35)	1.03 (0.72-1.48)	1.00 (0.73-1.37)
	Secondary	9580	26188	7.4	1.14 (0.91-1.43)	1.20 (0.86-1.69)	0.96 (0.71-1.28)
	Post Matric	811	2178	2.3	0.35 (0.14-0.86)	0.29 (0.71-1.19)	0.37 (0.12-1.127)
Marital status of mother	Never married	6320	17910	9.2	1	1	1
	Currently married	13844	41227	5.6	0.61 (0.49-0.75)	0.64 (0.48-0.86)	0.70 (0.52-0.94)
	Widowed / divorced	2822	7318	6.6	0.71 (0.51-0.99)	0.66 (0.39-1.11)	0.94 (0.52-1.42)

103

Annex 5.1 continued

Factor		Number of subjects	Person Years at Risk (PYAR)	Rate (/1000 PYAR)	RR (95% CI) 0-5 yrs	IRR (95% CI) 0-1 yrs	IRR (95% CI) 1-5 yrs
Mothers status at census after cohort entry	At home	26968	72464	6.6	1	1	1
	Migrant	3179	7207	4.7	0.71 (0.50-1.00)	0.88 (0.52-1.49)	0.70 (0.44-1.11)
	Died	500	1249	21.6	3.27 (2.16-4.96)	6.96 (4.16-11.64)	1.68 (0.84-3.36)
Number of female adults living in the home	1	8749	22609	5.4	1	1	1
	2	7726	20651	6.8	1.25 (0.98-1.61)	1.05 (0.74-1.48)	1.40 (0.98-2.00)
	3	6537	17285	7.6	1.40 (1.09-1.81)	0.99 (0.69-1.41)	1.76 (1.23-2.52)
	4 or more	7688	20446	8.9	1.64 (1.29-2.09)	1.27 (0.91-1.77)	1.83 (1.30-2.57)
Number of other children in the home	0	8639	21766	6.6	1	1	1
	1	11030	29450	6.6	1.00 (0.81-1.24)	0.92 (0.69-1.24)	1.13 (0.82-1.56)
	2	5937	15855	7.3	1.11 (0.86-1.42)	0.91 (0.64-1.30)	1.37 (0.95-1.97)
	3 or more	5127	13945	8.8	1.33 (1.03-1.73)	1.13 (0.78-1.64)	1.68 (1.18-2.39)
Antenatal clinic	Yes	13558	42172	10.3	1	1	1
	No	277	880	17.1	1.66 (0.99-2.77)	1.54 (0.77-3.10)	1.83 (0.86-3.92)
Delivered in a health facility	Yes	9991	30570	10.3	1	1	1
	No	3502	11098	11.4	1.11 (0.90-1.37)	1.02 (0.76-1.35)	1.28 (0.94-1.76)

Annex 5.2 Selected maternal characteristics by gender and nationality of household head

Characteristic	Male-headed (70.7%)	Female-headed (29.3%)	p-value (chi-square test)	Mozambican-headed (66%)*	South African-headed (32%)*	p-value (chi-square test)
Mother's characteristics						
Age 12-19	17.8%	22.9%	***	17.0%	20.6%	***
Age 20-29	42.5%	38.3%	***			
Age 30-39	22.7%	17.7%	***			
Age 40-49	17.0%	21.1%	***	16.7%	18.7%	***
No education	34.4%	23.4%	***	59.2%	14.1%	***
Secondary+ education	30.2%	42.0%	***			
Never married	20.7%	46.4%	***	20.6%	26.7%	***
Married	73.3%	23.6%	***			
Marriage dissolved	6.0%	30.0%	***			
Migrant	8.4%	16.0%	***	7.3%	12.3%	***
South African household head	57.0%	77.0%	***			
Mozambican household head	43.0%	23.0%	***			
Female-headed household				15.6%	31.9%	***
Not delivered in a health facility				43.8%	14.7%	***

*** p<0,000

* Nationality of household head had missing values on 2% of households.

Chapter 6

Maternal Vulnerability and Socio-economic Inequalities in Child Mortality in West Africa: An Exploratory Study

Morten Sodemann, Amabelia Rodrigues, Jens Nielsen and Peter Aaby

Summary

The aim of this research is to identify alternative socio-economic indicators, and examine these with regard to risk of child mortality. A further aim is to test the hypothesis that indicators of social capital and/or psychological health beliefs can in part explain variations in mortality risk independently, and to estimate if and to what extent 'favouritism' can explain variations in mortality risk in the study area.

The analysis outcome is the association between child mortality and socio-economic-, behavioural-, care related- and psychological factors. Three group interviews were performed with the aim of identifying culturally based indicators of wealth and health beliefs and testing of probable alternative indicators proposed by the researchers. The three groups (each with 6 persons) were comprised one of adult men, one with mothers and one with field assistants. Also a case-control interview study was performed with mothers who lost a child before the age of three years in the study area. The interview consisted of three different sections: one section with 70 statements with which the mother could express her level of agreement. In the second section the interviewer rated the appearance of the household and house surroundings. In the third section the interviewer gave an immediate impression of the mother through a choice of 34 personality descriptions.

Socio-economic status was estimated by quintiles of scores of the first component in a principal components analysis. The population studied via the wealth quintiles was all children less than 3 years of age resident in the study area.

There was no difference in actual knowledge among cases and controls. Traditional beliefs in the influence of spirits and witchcraft on childhood illness were equally frequent among cases and controls. No difference between cases and controls for the issue of maternal attention aspect was detected. It was shown that the degree of resignation in terms of what can be done in case of severe childhood illness depended strongly on ethnic group and for some of these it also depended on whether the mother had actually experienced a childhood death or not. External income sources were not reliable and cannot serve as socio-economic indicators in the present form. Among this particular group of poor mothers, favouritism was more important both in terms of beliefs and in actual possibility to make use of a relative or friend in the health care system. Our findings suggest that some of the alternative indicators including personality issues act more strongly among the poorest of the poor. It is our belief that some of our observations in larger samples could turn out to be independent factors in explaining mortality risk differentials among the poorest in developing countries.

Background

There is a clear sense among demographic health surveillance researchers that the present indicators of social and economic status are insufficient or indirect and sometimes not applicable in a developing country setting where the majority of the population live under the same poor conditions. Yet we observe significant differences in mortality within this group of similarly poor people. Our ideas of child mortality in developing countries are simplistic and its distribution has often been misunderstood because of insufficient attention to its context. High rates of child mortality in developing countries have variously been attributed to child neglect, economic scarcity, cultural traditions of child care, population pressure, low maternal education levels, lack of medical care, and insufficient basic resources. The relation between maternal education and child mortality is undisputable but there are indications that the influence is strongest in very low-income settings and furthermore health interventions may have a greater beneficial impact for less educated mothers (1-3). There are countries in which primary health services are so weak that they have no effect on the health of mothers and children; there are also countries in which health services tend to accentuate educational disparities because of differential access. Even the most well off and well educated families in developing countries experience infant mortality risks of 40-70/1000 deaths per 1000 live births. In the Bandim study area only 7% of the mothers who lost a 0-2 year old child from 1996-2001 were responsible for 34% of all deaths of children in that age group. (Unpublished data). At the maternity ward in Bissau, Guinea-Bissau there are two fold differences in the risk of caesarean section (and stillbirth) between ethnic groups (Unpublished data). The question is which factors make one poor mother capable of bringing up 7-8 children without loosing any while her equally poor neighbour is left with only three of her 8 children.

Demographic Health Surveillance sites (DHS) often deal with unexplained ethnic differences and significant changes in child mortality (CM) within small distances of the same area. More detailed studies of socio-economic indicators often come up with information that is useful in understanding the deeper mechanisms of mortality. A longitudinal study in the Bandim site of household determinants of death in multivariate analysis found that bed crowding of children less than 5 years of age increased mortality – *regardless of* maternal school education or possession of items. In the same study having pigs increased mortality by 10% even though pig owners took better care of their children with respect to vaccination (4). There are often numerous 'unofficial' ways of supplementing an otherwise unsatisfactory salary in developing countries and the existence of a 'natural' economy makes a concept like income inappropriate. Likewise it could be speculated that in societies where access to crucial health care is difficult and where family and ethnic relations are important elements when seeking health services, factors that affect CM may differ from factors that depict wealth. There has to be a certain level of knowledge of ways to secure child survival to allow individuals to escape from the circumstances imposed by broader socio-economic forces. Health equity is also about access and in many developing countries knowing a person in the health sector facilitates access to consultation, lab tests and medicine and can be more important than school education or a reasonable income. This is also termed 'favouritism', while it's relation to CM has never been investigated.

As CM is a major outcome in most demographic surveillance sites we have to look more closely at the process leading to the death of a child to find more satisfactory indicators. CM continues to be high in areas where there is easy access to health care. Several urban studies have demonstrated that 70-95% of children in developing countries are actually seen by a medical person before death and only 20-35% of these are hospitalised and eventually die in hospital (5-7). Most studies find no socio-economic difference in choice of first consultation among infant deaths (8). Factors that influence the choice of a given provider are maternal perceptions of the cause of illness, cost of care, distance from provider, availability and accessibility of provider, and past experience with a given provider (9, 10). A case-control study of childhood mortality risk including a wide range of child, maternal and socio-economic characteristics along with several indicators of access to health services, previous illness experiences and actual process of care fatal illness, demonstrated the *process of care* to be the strongest independent determinant of childhood death regardless of maternal schooling and other socio-economic indicators (11). Under these circumstances a mother can be extremely poor and still be able to handle her sick child rationally because she has autonomy and a strong social network. A rational process of care includes disease and severity recognition, reasonable home-medication, access to health care and ability to obtain contact with a health worker and perseverance follow-up until health has been regained. Women's autonomy in resource allocation is gaining increasing interest: greater freedom of movement, own income, say over purchases or number of children increases the likelihood of her choosing higher levels of antenatal care and likelihood of using safe delivery care (12,13). The extent of socio-economic 'buffer' and co-operating capacity of neighbours and friends independently alters the autonomy of the mother and thereby alters the mortality risk of her children.

Newer techniques developed to predict health behaviour of poor Afro-Americans used for measuring maternal attitude towards own ability to cope with a childhood illness (control) or not (chance, belief in power of others) also deserve to be tested in a developing country with appropriate adaptation to local circumstances (14)

There is a need to investigate the interface between rational (maternal) behaviour and access to health care issues as well as alternative measures of wealth on one side and on the other side the better established set of socio-economic indicators (wealth and schooling) that we are accustomed to use in studies in poor countries. An additional set of risk indicators that worsen the effects of poverty should be identified. These indicators could potentially be used to define health interventions that are more effectively aimed at the poorest in the highest risk of childhood mortality

Objectives

1. Identify alternative socio-economic indicators, and test these against risk of CM together with well-known socio-economic indicators usually associated with CM.
2. Test the hypothesis that indicators of social capital and/or psychological health beliefs can in part explain variations in mortality risk independently.
3. Identify process of care related indicators and relate them to socio-economic measures.
4. Estimate if and to what extent 'favouritism' can explain variations in mortality risk in the study area.

Does favouritism, maternal characteristics and identified alternative socio-economic indicators have an especially high impact on maternal attitudes towards the health sector among the poorest families in the lowest socio-economic quintile?

Outcome

Association between child mortality and socio-economic-, behavioural-, care related- and psychological factors.

Methods

The study is a household interview study, carried out under the Bandim health surveillance system. A two-phased approach was applied:

First phase
1. Three group interviews with the aim of identifying culturally based indicators of wealth and health beliefs and testing of probable alternative indicators proposed by the researchers were carried out. The three groups (each with 6 persons) comprised one group with adult men, one with mothers and one with field assistants.
2. Pilot testing of applied version of maternal health belief questions together with questionnaire on alternative socio-economic indicators.

Second phase
1. A case-control interview study with mothers who lost a child before the age of three years in the study area between 20/5/1999 and 5/12-2000 and controls matched on place of residence (the suburban population of 50.000 in the study area is divided into 24 sub districts), mother's age, and parity. The interview consisted of three different sections: one section with 70 statements (see Annex 6.1) to which the mother could express her level of agreement. In the second section the interviewer rated the appearance of the household and house surroundings. In the third section the interviewer gave an immediate impression of the mother through a choice of 34 personality descriptions divided into 5 groups of personality dimensions (Annex 6.2). Interviewers were allowed

to assign any number of characteristics. Traditional socio-economic and health indicators (birth weight, vaccination status and anthropometrics) are collected routinely through the child health surveillance and census.

Analysis

Socio-economic status was estimated by quintiles of scores of the first component in a principal components analysis. The population used for the scoring was all children less than 3 years of age resident in the study area between 20th of May 1999 and 5th of December 2000 (7,814 children). The following indicators were included in the analysis: type of roof, electricity, television, type of toilet, number of rooms and mean number of persons per bed. Maternal response to statements were scored to one of four levels: 'A' if the mother agreed totally, 'B' if the mother agreed a little or felt that the statement could be true sometimes, 'C' when the mother disagreed completely and 'D' if the mother did not know (excluded from analysis).

Immediate characterisation of the mother by the interviewer was analysed both as individual personality characteristics (dichotomous variables) and as groups of personality dimensions with a score in each group calculated by summing the value of each individual variable. Significant individual personality characteristics were isolated by stepwise elimination (probability (p)≤ 0.1).

Odds ratios were estimated by conditional logistic regression and controlled for interviewer. Each group of indicators, i.e. traditional indicators, statements and psychological characterisations, were tested group wise for their ability to predict cases and to explain variation.

The relationship between indicators and the risk of belonging to the lowest socio-economic quintile (i.e. 'the poorest mothers') was estimated by relative risks adjusted for being a case or a control. This group comprised 61 mothers and because of the small number of mothers in this group, a significance level of p<0.15 was chosen for this specific analysis.

Results

Group Discussions

The discussion of wealth quickly turned out to be taboo in the sense that if you disclose wealth you are more prone to robbery. Men were more materialistic than women in defining wealth (cows, pigs, a car) while mothers looked more at house appearance, type of schooling, education and relatives abroad. There was general agreement that information on 'un-official' ways of wage or income supplementation would never be disclosed to us. Mothers generally felt they could identify themselves in terms of social capital as a type of wealth though not mentioned spontaneously by any of the members of the discussion group. The male group had difficulty under-standing the concept of social capacity and tended to express more control of power in the household than their female counterparts expressed. The group discussions only added a few new concepts (e.g. 'relatives abroad regularly send money' and 'having underage children look after younger siblings'), while the discussions were more useful in excluding difficult questions (e.g. 'decision making and intentions in case of a terminally ill child').

Case-control Study

Two-hundred-and-fifty children died during the study period. This period was just after the 1998-99 war in Guinea-Bissau and many had not yet returned. Other mothers migrated or travelled for extended periods. Therefore, only 136 case interviews and 162 control interviews could be completed. Hence, 136 matched case-control pairs were included in the analysis.

Traditional Socio-economic Indicators

The ability of socio-economic status, maternal education and ethnic group to differentiate between cases and controls is shown in Table 6.1. Besides ethnic group (Muslim) there were no major differences between cases and controls. There was no significant effect of maternal school education.

Table 6.1   Traditional background factors

Indicator	Cases	Controls	Matched analysis* Odds ratio
	136	136	(95%CI)
Social class:			
I	29	32	1.18 (0.47-2.97)
II	31	34	1.17 (0.48-2.84)
III	32	26	1.53 (0.62-3.76)
IV	25	21	1.52 (0.60-3.81)
V	19	23	(Reference)
			p = 0.84
Maternal school education:			
None	52	51	0.93 (0.52-1.66)
0-4 years	32	37	0.77 (0.40-1.50)
> 4 years	52	48	(Reference)
			p = 0.74
Ethnic group:			
Pepel	72	63	1.13 (0.58-2.20)
Manjaco	16	19	0.78 (0.35-1.75)
Muslim	10	20	0.45 (0.18-1.13)
Other	38	34	(Reference)
			p = 0.22

* Conditional logistic regression.

Statements

Significant statements are listed in Table 6.2. Please refer to Annex 6.1 for a complete list of statements. Indicators with significant and borderline significant (p<0.15) relation to mothers belonging to the lowest socio-economic quintile (the poorest mothers) are shown in Table 6.6.

Favouritism   None of the indicators of favouritism could distinguish between case and controls (q21, q22 and q23). Among cases 50.9% (60/136) and 49.1% (58/136) of controls knew somebody in the health sector at the time of the interview. In case of severe childhood illness 51.4% (70/136) of case mothers and 47.8% (65/136) of controls found it impossible to obtain consultation and treatment at the hospital without money (q20). The poorest mothers were more confident that it was important to know a health person at the hospital, which is confirmed by the fact that the poorest mothers were more likely not to know any health workers (Table 6.6).

Religion- belief in spiritual powers or God   Whereas cases and controls were equally afraid of the influence of spirits and witchcraft (q2-q8), cases tended to be more confident that they could prevent childhood illness even when the disease could be God's will (q17). While 29/136 cases and 29/136 controls believed spirits and feticheiros could bring illness almost twice as many were afraid of illnesses brought by feticheiros and spirits (q2, q3). The same picture was seen in a statement regarding illnesses caused by the possession of specific animals in children and fear of these (q4 and q5). There was no difference in this distribution between the poorest and the rest of the mothers.

Belief in chance or fate   Cases were more likely than controls to agree that the survival of a severely ill child was a question of fate – primarily in the sense that illnesses have their own unpredictable course (q15). In terms of luck 41.2% of cases and 44.9% of controls found that child health was a matter of the individual child's luck (q10). Mothers from the poorest group were significantly more likely to rely on health as a matter of luck than richer mothers (Table 6.6). In terms of possible maternal negligence behind child deaths it is worth noting that 43.4% of cases and 47.8% of controls agreed totally that if a child was destined to death they as mothers could do nothing and 46.3% and 44.9% respectively disagreed completely to the same statement (q16). There was no difference in this distribution between the poorest and the rest of the mothers.

Belief in own power   Cases tended to be more likely to lose faith in the survival of her sick child (q9). Otherwise there was no difference in belief in own power in terms of prevention of disease (q11), obtaining a consultation (q13). The vast majority of cases (88.2%) and controls (80.1%) agreed that to see a doctor at the hospital required a persistent and stubborn mother. Only 58.8% of cases and 55.2% of controls were confident that they would always be able to obtain a medical consultation if they felt it was necessary (q13). Mothers belonging to the poorest social group were significantly more confident they would obtain a consultation if they needed it (Table 6.6). Cases were not more likely than controls to blame the loss of child on the

Table 6.2   Significant statements

Statement	*	Cases	Controls	Matched analysis** Odds ratio (95%CI)	
		136	136	**Crude**	**Controlled for interviewer**
q9: A mother can lose faith in	A	74	88	0.78 (0.34-1.82)	0.77 (0.32-1.81)
the survival of her sick child	B	33	18	1.79 (0.72-4.44)	1.82 (0.72-4.63)
	C	21	22	(Reference)	(Reference)
	D	8	8	p = 0.06	p = 0.06
q15: Disease is disease – a	A	73	56	1.61 (0.92-2.84)	1.73 (0.95-3.16)
mother can do nothing about it	B	21	30	0.85 (0.42-1.74)	0.80 (0.37-1.70)
	C	41	49	(Reference)	(Reference)
				p = 0.10	p = 0.10
q17: If gods want it so, I cannot	A	73	58	0.64 (0.37-1.10)	0.61 (0.35-1.08)
prevent my child's illness	B	7	12	1.30 (0.48-3.50)	1.50 (0.53-4.24)
	C	50	60	(Reference)	(Reference)
				p = 0.10	p = 0.10
q42: My husband always makes	A	91	79	2.32 (1.12-4.82)	2.51 (1.17-5.37)
the decisions regarding the	B	28	26	2.13 (0.94-4.88)	2.41 (1.02-5.69)
health, nourishment and school	C	17	31	(Reference)	(Reference)
issues of our children and never				p = 0.06	p = 0.06
leaves any decision to me					
q48: I have a less than 10-year-	A	32	52	0.41(0.22-0.76)	0.38 (0.20-0.72)
old child that looks after my	B	2	3	0.46 (0.07-2.93)	0.36 (0.05-2.57)
smaller children because I have	C	101	80	(Reference)	(Reference)
to go to work.		1	1	p = 0.01	p = 0.01
q51: My neighbours will always	A	71	56	2.49 (1.30-4.79)	2.83 (1.40-5.73)
help me in case of illness	B	35	30	2.31 (1.11-4.81)	2.81 (1.23-6.43)
	C	30	50	(Reference)	(Reference)
				p = 0.01	p = 0.01
q53: In case of illness I will	A	12	17	1.48 (0.61-3.62)	1.49 (0.57-3.87)
have problems because don't	B	13	6	0.49 (0.18-1.30)	0.48 (0.17-1.33)
have anybody to look after the	C	103	105	(Reference)	(Reference)
other siblings				p = 0.01	p = 0.01
q56: Children less than 3 years	A	25	33	0.81 (0.46-1.44)	0.75 (0.41-1.38)
of age I give cloroquine weekly	B	19	6	4.05 (1.35-12.14)	4.12 (1.36-12.6)
	C	92	97	(Reference)	(Reference)
				p = 0.01	p = 0.01
q58: Children less than 3 years	A	15	22	1.38 (0.69-2.74)	1.47 (0.73-2.97)
receive vitamins regularly	B	7	2	0.32 (0.06-1.55)	0.32 (0.06-1.62)
	C	114	112	(Reference)	(Reference)
				p = 0.01	p = 0.01
q62: All children less 3 years	A	62	78	0.61 (0.36-1.03)	0.55 (0.31-0.97)
sleep under same bed net	B	7	4	1.28 (0.36-4.60)	1.12 (0.30-4.19)
	C	63	50	(Reference)	(Reference)
	D	4	4	p = 0.13	p = 0.09

*   A: Agree totally; B: Agree to some extent/Sometimes could be true; C: Disagree totally; D: Don't know (excluded from analyses).

** Conditional logistic regression.

mother (q18). In fact 81.6% (111/136) cases and 80.9% (110/136) of controls disagreed completely with the statement.

Belief in power of the health sector   None of the indicators of faith in the health sector distinguished between cases and controls. It is interesting to note that 69.1% (94/136) of cases and 66.2% (90/136) of controls agreed somewhat or totally to the statement that many diseases could not be treated at the hospital (q18). At the same time 63.2% of cases and 57.4% of controls were very confident that a paediatrician at the hospital would be able to cure their sick child (q12). The poorest mothers were significantly less likely to choose private clinics as first choice in case of child illness and tended not to use pharmacies as first choice for advice and treatment (Table 6.6).

Disease prevention and management   Cases differed from controls in terms of more occasional than regular administration of weekly chloroquine (q56), more regular than occasional vitamin administration (q58), though associations were not very strong. Children of case mothers were less likely to always sleep under a bed net (q62). The poorest mothers were less likely to have chloroquine in the house, less likely to use anti-mosquito spray or mosquito coils, less likely to let their children sleep under a bed net and tended to be less likely to let a sick child drink and eat any type of foods (Table 6.6).

There was interaction between being a case and the ability of the poorest mothers to borrow medicine from neighbours: the poorest case mothers were the only mothers that anticipated problems borrowing medicine from neighbours.

Vulnerability   Cases more often stated that they anticipated definite care taker problems in case of child illness (q53). Cases were significantly likely than controls to let a less than 10-year-old look after younger siblings (q48). Cases complained more often that their husbands interfered too much in the care of their children (q42). Contrary to this cases were confident that neighbours would always help (q51). The poorest mothers were more likely to be short of money in the rainy season compared to richer mothers, more likely to rely on children to look after their siblings, less likely to be able to borrow medicine from their neighbours and less often received help or advice from their family (Table 6.6).

Alternative methods of income   Having a family member in Portugal or France was not associated with being a control and even among those that had such a relative there was no significant difference in the proportion of cases and controls that received money or goods from them regularly. There was no difference in this distribution between the poorest and the rest of the mothers.

Ethnicity   Though ethnic group was unable to distinguish cases from controls there are important observations: The tendency to let destiny determine the outcome of a childhood disease including death was significantly influenced by ethnic group. Mothers belonging to the Pepel group were more likely to let a child die if this was the destiny of the child, RR = 1.49 (1.17-1.90), adjusted for the mother being a case (q16). In the case of Muslim mothers their opinion depended on whether they had lost a child or not: case Muslim mothers were unlikely to let destiny decide RR = 0.27 (0.06-1.24), while control Muslims tended to let destiny decide RR = 1.59 (0.71-3.60). In terms of equity, Pepel mothers were more likely to belong to the poorest group of mothers RR = 1.68 (1.06-2.66) and Muslim mothers were more likely not to belong to this group RR = 0.42 (0.14-1.25).

Maternal Health

Self-assessed health   Self-assessed health was equal among cases and controls. The poorest mothers tended to feel more un-well than richer mothers RR = 0.72 (95%CI: 0.48-1.10).

Maternal mid-upper-arm-circumference (MUAC)   Overall there was no difference in MUAC between case and controls. However, ethnic group interacted significantly in the sense that among Muslims, case mothers had a higher MUAC (mean MUAC = 293.0 (95%CI: 263.4-322.6) than controls (mean MUAC = 258.8 (95%CI: 231.8-285.9)). An opposite tendency was observed among non-Muslim mothers (MUAC = 278.2 for cases and 287.8 for controls, p = 0.15)). There was no difference in maternal MUAC between the poorest mothers and other mothers (p = 0.88).

Table 6.3   Interviewers' immediate impression of the mother

Psychological Appearance		Case	Controls	Multivariate analysis* OR (95%CI)
Dedicated	Yes	45	52	0.58 (0.31-1.06)
	No	91	84	(Reference)
Rich or wealthy appearance	Yes	5	13	0.27 (0.08-0.95)
	No	131	123	(Reference)
Satisfied with her life and living conditions, no need to improve situation	Yes	16	6	6.73 (1.99-22.8)
	No	120	130	(Reference)
Silent person of few words	Yes	33	26	2.02 (0.99-4.13)
	No	103	110	(Reference)
Weak, powerless	Yes	9	3	3.66 (0.82-16.4)
	No	127	133	(Reference)
Timid, easily frightened	Yes	9	4	4.13 (1.04-16.4)
	No	127	132	(Reference)
Careless, sloppy	Yes	3	9	0.20 (0.03-1.17)
	No	133	127	(Reference)

* Conditional regression, stepwise elimination, adjusted for interviewer.

Interviewers Characterisation of Household and Surroundings

Cases tended to be less likely to be living under extremely poor conditions, OR = 0.68 (95%CI: 0.41-1.15), but otherwise none of the indicators of hygienic status, symbols of wealth of the household or signs of ability to locate time to 'house cosmetics' in the surroundings could significantly distinguish cases from controls.

Interviewers Characterisation of Mother

Individually significant psychological characteristics assigned by the interviewing assistant are shown in Table 6.3. Seven indicators could significantly predict cases. It is assumed that the immediate impression of the mother on the interviewer can be regarded as a proxy for either (1) how the mother generally appears to strangers (i.e. Health workers) or (2) how care seeking active she is or (3) How capable she is to overcome obstacles in the health sector. Taking this assumption into account mothers who lost a child was significantly more silent, weak, timid, satisfied and in lack of ambition compared with mothers who never lost a child. Results of the multivariate analysis of the grouped characterisations are shown in Table 6.4. Only one psychological group could significantly predict cases (group D).

The poorest mothers more often gave the interviewer a more frightened, timid and powerless impression than mothers from other socio-economic groups.

Comparison of Different Models to Predict Cases

To determine if the alternative indicators for socio-economic status had the ability to compete with traditional indicators of status, we analysed goodness-of-fit of each of the following models by their ability to explain variation and to predict cases: traditional indicators, significant statements, all psychological groups, psychological group D and individual significant psychological statements. The significant statements are the best models; psychological groups might be used.

Discussion

This study should be seen as a pilot study and there are methodological problems that need attention.

First of all we introduced a bias of information in as much as case mothers already *had* lost a child. This is an event that will probably change both the attitude and the behaviour of any mother. Being a pilot study the results presented here can therefore be nothing more than a mixture of the factors that actually lead to the loss of her child as well as other factors that changed because of the loss. The study should be seen as an inspiration for prospective studies incorporating some of the findings.

Secondly it was difficult to do the matching of cases and controls by parity age and area. Finding a mother with 7-10 live children is a rare event. At the same time the death of a child is a social event that is often followed by some degree of family disruption and subsequent migration. Because of limitations of the study this resulted in a sample that was too small, which is probably why so many associations are

Table 6.4   Groups of psychological characteristics

Psychological characterisation group	Score	Cases 136	Controls 136	Matched analysis* Odds ratio (95%CI) Crude	Controlled for interviewer
Psychological group A	0	20	21	1.20 (0.49-2.94)	1.27 (0.51-3.19)
	1	58	56	1.32 (0.62-2.82)	1.33 (0.61-2.90)
	2	41	38	1.39 (0.61-3.17)	1.43 (0.62-3.31)
	3	17	21	(Reference)	(Reference)
				p = 0.88	p = 0.87
Psychological group B	0	32	37	1.52 (0.54-4.28)	1.57 (0.54-4.63)
	1	67	60	1.90 (0.75-4.84)	1.97 (0.72-5.39)
	2	29	25	2.01 (0.72-5.63)	2.04 (0.69-6.01)
	3	8	14	(Reference)	(Reference)
				p = 0.47	p = 0.51
Psychological group C	0	41	47	0.62 (0.20-1.92)	0.57 (0.18-1.82)
	1	53	52	0.71 (0.24-2.06)	0.69 (0.24-2.03)
	2	19	22	0.62 (0.20-1.93)	0.55 (0.17-1.77)
	3	13	8	1.18 (0.28-5.03)	1.00 (0.23-4.40)
	4	10	7	(Reference)	(Reference)
				p = 0.67	p = 0.68
Psychological group D	0	98	93	0.50 (0.17-1.47)	0.48 (0.16-1.46)
	1	26	37	0.32 (0.10-1.03)	0.25 (0.08-0.86)
	2	12	6	(Reference)	(Reference)
				p = 0.11	p = 0.05
Psychological group E	0	56	49	1.14 (0.46-2.79)	1.33 (0.51-3.46)
	1	66	73	0.92 (0.39-2.18)	1.00 (0.41-2.42)
	2	14	14	(Reference)	(Reference)
				p = 0.67	p = 0.56

* Conditional logistic regression.

strong but only borderline significant. With a larger sample we might eliminate this problem.

Thirdly it seems as if a lot of variation was hidden by the matching (parity, area and age). The size of the study does not permit interaction tests with the matching variables against the outcome variable but it is obviously something that should be done in future studies.

Fourthly it was anticipated from the researchers that four options (Agree totally, agree a little, don't agree and don't know) for the statements would be enough but after the analysis had been completed it seemed clear that more variation among groups would have been demonstrated had we included more options (Such as: agree totally, agree a lot, agree to some extent, not entirely true, disagree completely and don't know).

The majority of the interviewer assigned characteristics of mothers showed no variation between case and controls while a selected number of characteristics yielded significant variation. As there were 3 field assistants trained to do this special interview, some degree of interviewer bias here is to be expected and cannot be avoided. We tried to carry out the interviews interviewer-blinded but this turned out to be impossible mainly because some of the questions invariably led the mother to mention her child loss.

The fairly strict matching had the effect that except for a non-significant positive effect of maternal education and a strong significantly positive effect of being a Muslim there was no association between being a case and traditional socio-economic indicators.

There was no difference in actual knowledge of a health person among cases and controls. Favouritism can of course still be a real problem in the care seeking. The question is how favouritism acts in the setting of severe childhood illness. The fact that cases were more concerned supports the notion that there is a problem and that it could be the cause of delayed or inappropriate care seeking. It could also be that our questions did not capture the information we request: type or rank of health person known, type of contact, actual presence of health person at hospital, prior or customary use of the person etc. Matching could also have obscured relations because acquaintance with a health care related person is both related to social level and to parity. Later studies are presently looking into this aspect and preliminary analyses indicate that what really matters in terms of child survival are the type of health worker the mother knows.

Traditional beliefs in the influence of spirits and witchcraft on childhood illness were equally frequent among cases and controls. Cases though tended to be more confident they could evade God's will in case of child illness. The duality of

Table 6.5   Comparing different models

Model	Ability to explain variation Pseudo-R^2	P(predict case | case)
Interviewer	0.0098	0.5066
Social class, ethnicity and mothers schooling	0.0336	0.5227
Questionnaire (significant statements)	0.4278	0.7434
Psychological factors (all groups)	0.0859	0.5560
Psychological factors (group D)	0.0419	0.5282
Psychological factors (outstanding factors)	0.1620	0.6018

urbanisation of traditional beliefs is depicted by the fact that twice as many mothers that believe in spirits and witchcraft as causes of illness are in fact afraid of spirits and witchcraft.

Probably due to a difference in experience cases were more confident than controls that they could do very little to avoid severe childhood illness. This duality in perception of disease causes is distinctive and probably depicts that mothers do not very often regard disease only as an act of God's will but they rather tend to see it as a conglomerate of repetitive illness episodes, fate, spirits, bad motherhood and low health quality. But in the end, fate and chance seem to play a surprisingly important role in perception of illness equally among cases and controls as nearly half in each group agreed that illness outcome and health were matters of luck.

The controversial issue of maternal negligence as an important factor in child mortality was addressed in the question on the possibility of a mother seeing death as her child's destiny. No difference between case and controls in this aspect was detected. But it is interesting that the question split mothers in two distinct groups: one half agreed totally to the statement and the other half disagreed totally leaving very few mothers without a clear idea on the issue. It was shown that the degree of resignation in terms of what can be done in case of severe childhood illness depended strongly on ethnic group and for some of these it also depended on whether the mother had actually experienced a childhood death or not.

Cases agreed very strongly to the statement that their husband was always interfering with the childcare and never left them to decide on their own. In terms of intervention possibilities it is worth noting that cases were more likely (even after having lost at least one child) to let a child less than 10 years of age look after younger siblings. A case also more often than controls anticipated problems finding a caretaker in case of child illness and this effect was significantly stronger among the poorest mothers.

The value of the non-blinded interviewer rating of the mother is still to be evaluated. The finding further elucidates the complexity of mortality differentials, wealth and vulnerability within homogenously poor communities. The influence of women's autonomy and vulnerability on care and health behaviour appears to be as important as other known determinants such as education. This is supported by findings from other developing countries (13, 15).

External income sources were not reliable and cannot serve as socio-economic indicators in the present form. We need to do more qualitative research in this field incorporating other alternative incomes (unofficial fees, family members employed in customs department, banks, ministries etc.).

The effect of having a supporting relative in Europe was insignificant. The general impression of the interviewers was that the mothers did not rely on this support and they never knew when the help would come.

Maternal health status measured by MUAC showed an interesting interaction with ethnic group in the sense that Muslim case mothers had higher MUAC than Muslim controls whereas the opposite was the case with non-Muslim cases although less pronounced. This needs to be explored further in existing longitudinal datasets from the Bandim site with these measures.

Interviewer impressions of the house surroundings, household appearance and maternal characteristics showed very little variation except for determination-independent index where cases scored highest. Though borderline significant the difference was small and probably not interesting in terms of childhood mortality interventions. What is more interesting is that cases did not give the interviewers an impression of shyness or timidness but rather of being weak or powerless. This is interesting in terms of care seeking and obtaining relevant medical care in due time for severe childhood illness.

Our simplistic psychological profile of mothers to identify 'weak' and 'strong' mothers just by the immediate impression of the mother was effective but could probably yield a lot more detail with the assistance of a psychologist. Under the assumption that an immediate subjective impression predicts caring abilities of a mother the findings of this study could theoretically explain mortality differentials among the poorest: mothers who lost a child were more timid and powerless while satisfied with their conditions and with no apparent need to improve their situation.

Issues of equity were addressed through a focus on mothers belonging to the poorest socio-economic group. Though we were unable to demonstrate significant differences between cases and controls, a number of observations are important for future longitudinal studies.

Among this particular group of poor mothers, favouritism was more important both in terms of beliefs and in actual possibility to make use of a relative or friend in the health care system. The poorest were generally left only with the option of the public health service and were less likely to make use of alternative places of consultation like private clinics or pharmacies. The poorest mothers did not differ from richer mothers in their traditional beliefs but tended to rely more on luck than other socio-economic groups. Mothers in the lowest quintile The poorest mothers did significantly less to prevent illness as measured by their reluctance to use mosquito barriers e.g. bed nets or coils while they were much more likely to give chloroquine regularly to their children. The latter could be a risk factor but could also be a decision made by the mother after having lost one child. Cases were more often short of money in the rainy season – a period of time when petty traders have difficulties because of transport problems and less food output from their farming. No personality issues applied specifically to the poorest mothers.

Some important observations were made in terms of disease management: the poorest mothers were less likely to let the child eat and drink at leisure. This could potentially be harmful to severely ill child and needs further investigation. In case of child illness the poorest mothers also had to rely much more on under age caretakers, got no help from the family, and did not count on an ability to borrow medicine from neighbours. They also complained that husbands were interfering with the care of the child and did not leave decisions in their hands – this issue would be an interesting issue for anthropological studies. A recent Brazilian study found that the risk of child malnutrition was elevated in families where the mother had 3 of 4 screening symptoms of depression (16). It is important to include such variables both as risk factors for- and consequences of child death.

In conclusion, traditional socio-economic indicators were of less importance probably due to the strict matching on sub district and parity. The matching probably also obscured the significance of favouritism. Apart from case mothers being aware of the fact that they might give up on a very sick child there were no major differences in attitudes towards sick children. Three statements were so strongly associated with

Table 6.6   Alternative indicators associated (p<0.2) with very low socio-economic status (N = 61 mothers). Relative risk adjusted for being a case

Indicator	Risk of being in lowest socio-economic quintile Relative risk (95%CI)
Maternal characteristic:	
Timid, easily frightened	0.29 (0.04-2.17)
Weak, powerless	2.47 (0.81-7.51)
Agrees totally with statement (Yes vs. no):	
'Often has a daughter less than 10 years of age to look after younger siblings'	1.38 (0.95-2.02)
'It is important to know a health person'	1.67 (1.00-2.89)
'I know a health worker'	0.66 (0.44-0.99)
'Child health is a matter of luck'	1.36 (1.02-1.81)
'If I want to I will always obtain a consultation'	1.37 (1.12-1.67)
'I always have chloroquine in the house'	0.54 (0.31-0.96)
'I use anti-mosquito spray regularly'	0.48 (0.23-1.00)
'I use mosquito coils regularly'	0.52 (0.21-1.28)
'My children always sleep under a bed net'	0.67 (0.41-1.10)
'In case of illness my child can eat and drink anything'	0.55 (0.26-1.16)
'We never have money in the rainy season'	1.36 (0.96-1.94)
'In case of illness I can always borrow medicine from my neighbours'	**Crude**: 0.67 (0.39-1.15) **Cases**: 0.39 (0.15-1.07)* **Controls**: 0.96 (0.48-1.94)*
'My family always helps me with advice in case of illness'	0.82 (0.64-1.00)

* p (interaction) = 0.1.

being a case that they deserve further investigation: case mother's complained over husbands interfering, had to rely on very young care-takers and were not confident they could get necessary help from their family when in need. The latter was significantly stronger among the poorest cases. Our findings suggest that some of the alternative indicators including personality issues act stronger among the poorest of the poor. It is our belief that some of our observations in larger samples could turn out to be independent factors in explaining mortality risk differentials among the poorest in developing countries and they will most likely lead to stronger and more effective health interventions for the poorest of the poorest.

The present findings should clearly be investigated in longitudinal studies of birth cohorts after refinement of statements and the psychological characteristics.

References

Adams, A.M., Madhavan, S. and Simon, D. (2002). 'Women's social networks and child survival in Mali'. Soc Sci Med; 54(2):165-178.

Bentley, M.E. (1988). 'The household management of childhood diarrhea in rural north India'. Soc Sci Med; 27(1):75-85.

Bloom, S.S., Wypij, D. and Das, G.M. (2001). 'Dimensions of women's autonomy and the influence on maternal health care utilization in a north Indian city'. Demography; 38(1):67-78.

Caldwell, J.C. (1993). 'Health transition: the cultural, social and behavioural determinants of health in the Third World'. *Soc Sci Med*; 36(2):125-135.

Cleland, J.G. and van Ginneken, J.K. (1988). 'Maternal education and child survival in developing countries: the search for pathways of influence'. Soc Sci Med; 27(12):1357-1368.

'Crowding and Health in Low-Income Settlements'. Case Study Report, Bissau. 1995. Copenhagen, Cowi.

de BL Carvalhaes, M.A. and D'Aquino Benício, M.H. (2002). 'Capacidade maternal de cuidar e desnutricão infantil'. Rev Saúde Pública; 36(2):188-97.

Gutierrez, G. (1994). 'Study of the Disease-Health Seeking-Death Process: Another ise of the verbal autopsy'. In Reyes, H., Martinez, H., Tome, P. and Guiscafre, H., (eds). International Journal of Epidemiology 23, 427-428.

Hindin, M.J. (2000). 'Women's power and anthropometric status in Zimbabwe'. Soc Sci Med; 51(10):1517-1528.

Rosenzweig, M.R. and Schultz, T.P. (1982). 'Child mortality and fertility in Colombia: individual and community effects'. Health Policy Educ; 2(3-4):305-348.

Sodemann, M., Jakobsen, M.S., Molbak, K., Alvarenga, I.C., Jr. and Aaby, P (1997). 'High mortality despite good care-seeking behaviour: a community study of childhood deaths in Guinea-Bissau'. Bull World Health Organ; 75(3):205-212.

Sodemann, M., Veirum, J., Biai, S., Nielsen, J., Bale, C., Jakobsen, M. et al. (2003). 'Reduced case-fatality among hospitalised children during a war in Guinea-Bissau: a lesson in equity'. Accepted, Acta Paed.

Slutsker, L., Bloland, P., Steketee, R.W., Wirima, J.J., Heymann, D.L. and Breman, J.G. (1996). 'Infant and second-year mortality in rural Malawi: causes and descriptive epidemiology'. Am J Trop Med Hyg; 55(1 Suppl):77-81.

Terra de Souza, A.C., Peterson, K.E., Andrade, F.M., Gardner, J. and Ascherio, A. (2000). 'Circumstances of post-neonatal deaths in Ceara, Northeast Brazil: mothers' health care-seeking behaviors during their infants' fatal illness'. Soc Sci Med; 51(11):1675-1693.

Tinsley, B.J. and Holtgrave, D.R. (1989). 'Maternal health locus of control beliefs, utilization of childhood preventive health services, and infant health'. J Dev Behav Pediatr; 10(5):236-241.

Villa, S. (1994). 'Muertes en el hogar en niños con diarrea o infección respiratoria aguda después de haber recebido atención médica'. In Guiscafre, H., Martinez, H., Urban, J.C., Reyes, S., Lezana, M.A. and Gutierrez, G. (eds). Bol Med Hosp Infant Mex 51, 233-241.

Annex 6.1   Statements

Q1 My children are always well
Q2 Diseases can be caused by *feticheiro*, spirits
Q3 I am afraid of *feticheiro*, spirits
Q4 Wolves, cats or monkeys taking possession of a child can cause childhood diseases
Q5 I am afraid of diseases caused by wolves, cats or monkeys taking possession of child
Q6 Diseases can be caused by bad milk
Q7 In case of *ceremonia*, witchcraft or bad milk I can do nothing to prevent disease
Q8 In case of *ceremonia*, witchcraft or bad milk a physician can do nothing to prevent disease
Q9 A mother can lose faith in the survival of her sick child
Q10 Health is a matter of the child's luck
Q11 I can prevent illness if I want to
Q12 I am certain that a physician can cure my sick child
Q13 If I want to I will always obtain a medical consultation
Q14 Obtaining consultation takes a stubborn and persistent mother
Q15 Disease is disease – a mother can do nothing about it
Q16 If death of a child is its destiny I can do nothing
Q17 If gods wants it so I cannot prevent my child's illness
Q18 Only the mother can be blamed if her child dies
Q19 Many diseases cannot be treated at the hospital
Q20 Even when my child is severely ill: no money means no consultation or medicine
Q21 It is very important to know somebody at the hospital
Q22 If you don't know anybody at the hospital it is waste of time to seek help there
Q23 I personally know somebody in health
Q24 – And I used this person to obtain help
Q25 I never seek consultation at health centre because they don't behave well
Q26 I never seek consultation at health centre because they only give you a prescription
Q27 I never seek consultation at health centre because it is too difficult to obtain
Q28 I never seek consultation at health centre because I don't have a husband and I have no money and I have none to support me
Q29 I seek consultation at health centre because only nurses consult there
Q30 Private clinics are the best places to consult
Q31 Pharmacies with doctors are better places to consult
Q32 Paediatric ward outpatient clinic is only place to consult sick child
Q33 Always seek consultation at private clinic as first choice
Q34 Always seek pharmacies as first choice for consultation
Q35 In case of illness I always use my spare money and do not have to wait for my husband
Q36 In case of illness I always wait for my husband to come with the money
Q37 In case of illness I always await the counsel or decision of my husband
Q38 In case of severe illness I will always be able to get my child hospitalised

Q39 My husband will always help me with money

Q40 If I don't have money I can always arrange it

Q41 I never have money in rainy season

Q42 My husband always makes the decisions regarding the health, nourishment and school issues of our children and never leaves any decision to me

Q43 On many instances I did not by all the prescribed drugs because I did not have money

Q44 On many instances I did not by all prescribed drugs because I awaited an improvement in my child's condition

Q45 I always keep spare money for medical urgencies/disease of my children

Q46 In case of illness I often find that I don't know what to do

Q47 In case of illness my child can eat and drink anything

Q48 I have a less than 10-year-old child that looks after my smaller children because I have to go to work.

Q49 I have a less than 10-year-old child that looks after my smaller children in case of illness among them.

Q50 My family always helps me with advice in case of illness

Q51 My neighbours will always help me in case of illness

Q52 In case of hospitalisation I will have problems because I lack money

Q53 In case of illness I will have problems because don't have anybody to look after the other siblings

Q54 If I don't have any drugs I can always borrow from my neighbours

Q55 I always have chloroquine in the house

Q56 Children less than 3 years of age I give chloroquine weekly

Q57 I have vitamins in the house

Q58 Children less than 3 years receive vitamins regularly

Q59 I often treat my sick children with drugs that I keep in the house

Q60 I use anti-mosquito spray regularly

Q61 I use mosquito coils regularly

Q62 All children less than 3 years of age sleep under bed net

Q63 I myself I am always feeling ill

Q64 I am always travelling because I by and sell

Q65 I have relatives in Portugal/France/Europe

Q66 They always send me money

Q67 At least one bed with more than 1 child less than 3 years old

Q68 Children always sleep under bed net

Q69 Some children sleep on the floor

Q70 Domestic animals sleep in the same room as us

Annex 6.2 Psychological indicators

Group A
Strong, effective, capable, self confident, independent, determined.

Group B
Dedicated, satisfied, providing own needs.

Group C
Timid, modest, reserved, weak, not competent, not capable, quiet, disorganised, vulnerable, no faith in life.

Group D
Irrational, illogical, confused, not interested, self-remorsing.

Group E
Very religious, very happy, very rich personality.

Chapter 7

Parents' Socio-economic Status and Social Support as Risks for Child Mortality: Consideration of Health Equity in The Gambia

Amy Ratcliffe, Kate Halton, Rosalind Coleman, Maimuna Sowe and Gijs Walraven

Introduction

A matched case-control study of mortality to children under age five was conducted to consider associations with parents' socio-economic status and social support in the Farafenni Demographic Surveillance Site (DSS). Cases and controls were selected from Farafenni DSS, matched on date of birth, and parents were interviewed about personal resources and social networks. Parents with the lowest personal socio-economic status and social support were identified. Multivariate multinomial regression was used to consider whether the children of these parents were at increased risk of either infant or 1-4 mortality, in separate models using either parents' characteristics. There was no benefit found for higher SES or better social support with respect to child mortality. Children of fathers who had the poorest social support had lower 1-4 mortality risk (OR = 0.52, p = 0.037). Given that socio-economic status was not associated with child mortality, it seems unlikely that the explanation for the link between father's social support and mortality is linked to resource availability. Explanations for the risk effect of father's social ties may lie in decision-making around health maintenance and health care for children.

> (In the traditional African family system) ... co-operation is gained through a network of mutual obligation and the exchange of resources between kin. Investments are made in people, in relations of kinship and affinity, which can be called upon in time of need, ideally providing a permanent source of sustenance and security.
>
> Christine Oppong, 1992, p. 70.

Background

There are often issues of health equity between stratified groups within populations. These differences may be masked by reports of health indicators at the population level. The identification of these differences and the means to level them requires first, an identification of the relevant strata and second, a comparison of the health differences between them. This analysis is an attempt to consider whether such strata exist for families in the Farafenni Demographic Surveillance Site based on socio-economic status and social support with respect to child survival.

The multidisciplinary nature of public health allows researchers to consider child mortality in the context of the society, community and family. Putting child mortality in this context requires an understanding of the various pathways through which factors at these levels can affect child health. Following the Mosley-Chen framework for analysis of child survival in developing countries pathways through which external, indirect factors affect health can be elucidated (Mosley and Chen, 1984). Mosley and Chen identified a set of proximate determinants of child mortality:

- maternal factors – e.g. age, parity;
- environmental contamination – e.g. sanitation, crowding, pathogens;
- nutrient deficiency – e.g. protein deficiency, micronutrient imbalance;
- injury – accidental or intentional; and
- personal illness control – preventative strategies and curative health care.

Indirect factors such as cultural norms, family socio-economic status, or parent's behaviour only relate to child health and survival distally through these proximate determinants.

The challenge to public health is to identify which factors are relevant to child mortality and can be addressed in public health programmes. Many studies have focused on indirect factors in the context of the family. Parents, and mothers in particular, have received considerable attention. Epidemiological studies that include sociodemographic determinants and socio-economic status (SES) of families and parents have shown that poorer, less educated families tend to have higher mortality among their children in many populations. A family's ability to maintain health and to care for ill family members depends on their access to resources and effective use of these resources to address health needs.

Child mortality remains high in rural West Africa. The Farafenni Demographic Surveillance Site (DSS) reported an infant mortality rate of 74.3/1000 and 1-4 mortality of 40.2/1000 (Ratcliffe, 2002). Residents in the DSS are poor, rural farmers and the population is fairly homogenous with respect to sociodemographic factors. Three ethnic groups are represented but health differences between them are not striking. Previous work in this population has suggested that some individuals have relatively low social support (Ratcliffe et al., 2000). Where everyone is already poor, weak social support may leave certain families particularly vulnerable for reasons suggested below.

In rural West Africa, reciprocal relationships and social networks of support are important as a mode of insurance for times of need (Oppong, 1992). When resources are available it is expected that individuals will share among their kin and social allies. Reciprocating relationships facilitate co-operation in a population where most residents are living on very little. The ability to garner support from others for family health needs may be an important determinant of health maintenance and access to health care in these poor, rural communities.

While most individuals in rural West Africa are born into a network of social support in their village, some individuals are separated from that support through migration, family death or disruption, illness or impairment, or through a reputation of 'refusing advice'. In virilocal marriages, brides leave their natal homes and although they retain important ties to their own kin, their social alliances in the husband's community are developed over time and respect is earned through reproduction and labour (Oppong, 1992). Such migration may be especially difficult for women who marry far from their natal families, across national borders, or for those women who for one reason or another are unable to foster social relationships in their new community. In recent qualitative work on men's marriages and fertility, men who had poor relations with their kin and natal families were less optimistic about their future than other men in similar circumstances (Ratcliffe, 2000).

In developed countries, social capital has been identified as a mediator between socio-economic status and health for communities and populations (Kawachi and Kennedy, 1997). Social capital is described as the social organization of the community, the features that facilitate cooperation and mobilization (Putnam, 1995). Certain communities have been found to mobilize resources for health through cooperative networks even in the face of poor socio-economic status that might otherwise be a limiting factor. Like communities, certain families may be better able to provide for the health of their family members by calling upon their social networks for practical support, despite poor SES.

A family's ability to provide for health maintenance and to mobilise resources in times of need can be considered among the indirect determinants of child health. We hypothesise that children whose parents have limited personal resources and/or weak social support are at highest risk for mortality in the Farafenni DSS. In a case-control study design, parents with the lowest SES and the weakest social support were identified and associated risks for mortality among their children were considered.

Methods

Study Population

The population living in 40 rural villages near the town of Farafenni in the North Bank Division of The Gambia has participated in demographic surveillance conducted by the UK Medical Research Council since the early 1980s. By mid-1999 the population was 16,202 people of three main ethnolinguistic groups: Mandinka (43%), Fula (20%) and Wollof (36%). Villages range from 40 to 1,221 residents and are organized into compounds based on extended family with up to 149 residents and 18 on average. Leadership within the villages usually rests with the men at the village, compound and household levels. The population is almost entirely Muslim and polygyny is widespread. Use of routine health services, especially maternal and child health services is high, resulting in vaccine coverage greater than 80% and prenatal care coverage of 94%. Use of care for treatment for episodes of illness is not as good. The most common causes of child deaths are malaria, acute respiratory infections and diarrhoea (Ratcliffe et al., 2002).

Sample Selection

Cases and controls were selected from the Farafenni Demographic Surveillance Site (DSS) registry. All registered deaths that occurred between 31 March 1999 and 1 April 2001 to children aged less than five-years old were included as cases. Controls were surviving children matched on date of birth. Surviving twins were not eligible as their own twin's control. Siblings of cases were not excluded as controls matched for non-sibling cases. The full sample included 232 cases and 232 controls. Survival time was calculated as time from birth to death or censoring which occurred on 31 March 2001.

These cases and controls were linked to their parents through the DSS registry. 432 mothers had been registered within the DSS registry; 6 of these mothers had died and 10 had emigrated. 401 fathers had been registered within the DSS registry; 12 of these had died and 26 had emigrated. The target sample of parents therefore included 416 resident mothers and 363 resident fathers. 396 mothers and 337 fathers were successfully interviewed, producing response rates of 95% and 93%, respectively.

Data Collection

Mothers and fathers of both cases and control children were interviewed separately to complete a short questionnaire that included information on personal resources and social networks. Interviews were conducted from July through September 2001. Respondents were not aware of the selection criteria or their inclusion as parents of either cases or controls. Interviewers were also blinded to the status of participants as parents of either cases or controls. To ensure this, no information relating to the index child was included on the questionnaire and no questions related directly to child illness or deaths.

In addition to specific questions about sociodemographics, resources and relationships, respondents were asked to agree or disagree with statements related to support they could expect to receive from members of their social networks. Interviewers helped them rank their level of agreement from strongly disagree to strongly agree on a scale from 1 to 5. Four of these statements related to practical support from family or community. These statements are included below in their English translation:

- There is someone who lives in this village whom you would trust to help you solve a problem. *Give example: you need to put a new roof on your house.*
- If you were away, there is someone in this village whom you would trust to take care of your family just like you would. *Give example: a family member needs to go to the clinic.*
- In your absence, could you trust your husband/wife to make decisions about the children's care (*tapoto*) in the same way that you would.
- Is there someone from your birth family who would come if you called them in case of an emergency?

2000 Household Socio-economic Survey

Ending in June 2000, 1,169 household heads resident in the study area were interviewed to collect socio-economic data on the household. In a short interview that was conducted as part of the normal DSS surveillance rounds, individuals were asked about occupation, housing structure, possessions, farm outputs, livestock, and school attendance for resident children.

Description of Datasets

Data for this analysis was derived from four distinct datasets. Descriptive statistics for cases and controls (N = 464); interviewed fathers (N = 337); interviewed mothers (N = 396); and household heads who participated in the 2000 survey (N = 1169) were run from these four separate datasets. New variables and composite scores were created as described below.

These datasets were then merged to create the two working datasets, one that included 417 children matched to their mothers and another that included 375 children matched to their fathers. Similar proportions of cases and controls matched to their parents. Mothers and fathers were able to repeat in these datasets after merging to children when siblings were included among the set of cases and controls. Twenty-three mothers repeated with 9 matched to both a case and a control. Thirty-eight fathers repeated with 21 matched to both a case and a control.

Compound SES Score

A summed variable (ownership sum) was created for all households that participated in the 2000 Household SES Survey by adding the 0,1 dummy variables that indicated possession of certain items or household features: an iron or asbestos roof on the household head's house; a bicycle; a radio; and a manufactured bed.

The DSS registry does not list children by households so cases and controls could not be matched to their own household data. Household variables (ownership sum, number of cows owned, number of sheep or goats owned, and number groundnut sacks sold from the previous harvest) were averaged within compounds and merged with the data on children and their parents based on the child's residence at death or censoring. Principal components analysis (PCA) was used to determine appropriate weights for a composite (compound SES) of these averaged variables. A dummy variable was created to indicate where the child's compound SES score fell in the bottom quartile of the distribution of all compounds where data was available.

Parent SES Sum

A summed variable (parent SES) was created to capture the socio-economic status of the individual parents by adding the 0,1 dummy variables that indicated certain aspects of socio-economic status. Variables included were: possession of money; possession of goods to be sold; a possession that could be sold in an emergency; a possession that could be used to guarantee a loan; and uses cooking oil at least once a week. All of the abovementioned possessions were qualified as being within the personal control of the respondent. A dummy variable was created to indicate where parent SES sum fell below 3.

Parent Support Score

Principal Components Analysis (PCA) was used to create composites (parent support), with weights determined separately for fathers and mothers, to reflect parents' reported ability to rely on others for practical support in times of need. Variables included in this composite are based on the responses, scaled from 1-5, of agreement with the statements listed in the data collection section above.

For mothers, the first component from PCA (eigenvalue = 1.662, describing 42% of the total variance) was chosen. Weights for the mothers' parent support composite were: 0.764 (problem solver), 0.726 (family carer), 0.547 (husband cares for family), 0.502 (birth family come in emergency). For the fathers, the first component from PCA (eigenvalue = 1.559, describing 40% of the total variance) was chosen. Weights were for the fathers' parent support composite were: 0.743 (problem solver), 0.648 (family carer), 0.617 (wife cares for family), 0.496 (birth family come in emergency). As the coefficients for all four variables were positive for both mothers and fathers, individuals would score high if they could depend on good support with respect to all four aspects. A dummy variable was created to indicate where parent support scores fell in the bottom quartile for the distributions for mothers and fathers, respectively.

Comparison of Cases and Controls

Within each of the working datasets, children with mothers and children with fathers, comparisons were made between cases and controls with respect to the parents' sociodemographic indicators, compound SES, parents' socio-economic indicators, and indicators of parents' social networks. Tests for proportions were done using Chi-square values and corresponding p-values. Tests for means were done using analysis of variance (ANOVA) and F statistics with corresponding p values. Significance was considered at the 0.05 and 0.10 levels.

Multivariate Logistic Regression Models of Child Mortality

The aim of this analysis was to consider parent's socio-economic status and social support as health equity stratifiers within the population, controlling for other risk factors. Based on the hypothesis that the most vulnerable individuals would be those with the lowest SES and the weakest social support, dummy variables were used to indicate parent SES sum less than 3 and membership in the lowest quartile of parent support for mothers and fathers, respectively. A set of control variables were also considered for inclusion in these models. Based on a recognition that infant and 1-4 mortality may be associated with different indirect factors, multinomial logistic regression models were built to predict infant deaths and 1-4 deaths separately, using either mother's or father's characteristics, compared to controls.

Results

Descriptive Statistics of Households

In the 2000 survey of household heads (Table 7.1), 86% listed farming as their primary profession and 27% of those farmers did not sell any produce from their farms. The average gross income from the 1999/2000 groundnut harvest was between 2,632-2,961 dalasis (US $219-247). Of the household heads, 96% were living in houses with mud walls and 36% had thatched roofs. Important ethnic differences were found in SES indicators. Mandinka (84%) were more likely to have an iron or asbestos roof compared to both Fula (43%) and Wollof (49%). While Fula (50%) and Wollof (53%) are as likely to own cows compared to Mandinka (34%), Fula had larger herds on average (8.9 cattle) compared to Mandinka (4.6) and Wollof (4.2).

Descriptive Statistics of Cases and Controls

Cases (57%) were slightly more likely to have been boys than controls (54%). Among cases, the mean age at death was 1.06 years and 44.0% were less than one year old at the time of their death. There is evidence of a seasonal effect on mortality, with 74% of deaths during the malaria season from August to December. Compound SES scores for residences of cases and controls were not significantly different.

Table 7.1   Descriptive statistics 1,169 households, Farafenni DSS 2000 household socio-economic survey, percent reporting or mean as specified

	Percent
Household Head (N = 1169)	
Ethnicity – Fula	22.3
Mandinka	46.0
Wollof	30.8
Mean Age	54.0
Female	3.5
Possessions and Housing	
Roof – Iron/Asbestos	63.6
Thatch	36.4
Bicycle	11.4
Radio	64.2
Manufactured Bed	52.7
Mean Ownership Sum	1.9
Ownership Sum = 0	12.2
Household Production	
Mean # Cows (N = 478)	5.4
Mean # Sheep/goats (N = 646)	4.1
Mean # Groundnut Sacks Sold (N = 710)	16.7

Descriptive Statistics of Mothers and Fathers

The sociodemographic characteristics of 396 mothers are presented in Table 7.2. Mothers were on average 31 years old at interview with a range from 16 to 59. Eight percent of the mothers had attended formal schooling but this fraction was as high as 14% among the Fula mothers and as low as 3% for the Wollof mothers. Nearly all of the mothers were farmers but slightly less than half claimed this as their sole occupation. Fula mothers were most likely to have farming as their sole occupation (60%) compared to Wollof (53%) and Mandinka (35%). Among mothers who had another occupation, plaiting hair was the most common craft specified. Mothers had an average of 5.0 children (1-12) and 0.9 child deaths (0-4) each registered in the DSS registry.

The sociodemographic characteristics of 337 fathers are presented in Table 7.2. Fathers were on average 45 years old at interview with a range from 22 to 78. Five percent of the fathers had attended formal schooling but this was as low as 2% among the Wollof fathers compared to 7% for both Fula and Mandinka. Nearly all of the fathers were farmers and approximately a third gave this as their only occupation. Forty-two percent of the Wollof fathers and only 31% and 28% of the Fula and Mandinka fathers, respectively, gave farming as their sole occupation. The range of occupations for fathers was much more diverse than those of the mothers. Specific occupations were associated with ethnicity; Mandinka were more likely to be craftsmen or tradesmen while Fula were more likely to be herders or fisherman. Fathers had an average of 8.3 children (1-27) and 1.2 child deaths (0-7) each listed in the DSS registry.

Socio-Economic Status of Mothers and Fathers

Nearly half of the mothers had money in their possession at interview and two-thirds had something they personally could sell to raise money in an emergency (Table 7.2). Mandinka mothers (58%) were more likely than Wollof (40%) and Fula (31%) to have money in their possession. Parent SES sum for mothers ranged from 0-5 with a mean of 3.4 (median 3). This varied with ethnicity; Mandinka mothers had a higher mean sum (3.8), compared to Fula (3.1) and Wollof (3.2) mothers. Twenty-four percent of the mothers had an SES sum less than 3. Mothers with the lowest SES sums were also less likely to be a member of a village group (OR = 0.45, p = 0.040) and more likely to have farming as their sole occupation (OR = 3.28, p = 0.000). These mothers were similar to others with respect to residence in their home village, having family members living in their village, formal schooling, and English literacy.

Sixty-two percent of the fathers had money in their possession at interview (Table 7.2). As many as 75% of the Mandinka fathers and only 53% of both the Wollof and Fula fathers were in possession of money. Fathers reported themselves better off than mothers with regard to ability to raise money in times of need. Parent SES sum for fathers ranged from 0-5 with a mean of 4.0 (median 4). This varied with ethnicity; Mandinka fathers had a higher mean sum (4.2), compared to Fula (3.8) and Wollof (3.9) fathers. Ten percent of the fathers had a parent SES sum of less than 3. Fathers with the lowest SES sums were also less likely to have family members living in their village (OR = 0.32, p = 0.098); more likely to have farming as their sole occupation

Table 7.2   Sociodemographic and socio-economic characteristics of 396 mothers and 337 fathers, as percent reporting or mean as specified

	Mothers	Fathers
Frequency	396	337
Age	31.3	45.4
Education		
Attended Formal School	8.3	5.0
Literate in English	5.6	4.5
Attended Qu'ranic Training	82.3	93.8
Ethnicity		
Fula	21.7	21.4
Mandinka	39.4	38.6
Wollof	38.4	39.8
Occupation		
Farmer	99.2	96.4
Farming only occupation	47.7	34.1
Practices craft or trade	7.8	19.2
Practices trading or business	44.2	16.3
Marabout	–	6.8
Imam or Islamic leader	–	3.6
Fisher or herder	–	12.2
Professional	–	5.6
Other	–	2.1
Children		
Mean Number Children in DSS	5.0	8.3
Mean Number Child Deaths in DSS	0.9	1.2
Parent SES		
In possession of money	45.5	61.7
Could sell something in emergency	66.1	93.8
Owns something to guarantee a loan	85.1	95.9
Has goods or produce to sell	61.6	73.9
Uses oil for cooking less than once weekly	22.2	26.1
Mean Parent SES Sum	3.4	4.0

(OR = 2.49, p = 0.018). These fathers were similar to other with respect to residence in their home village, membership in village groups, formal schooling and English literacy.

Social Support of Mothers and Fathers

Less than half of the mothers were living in their home village, defined as the village where they spent their childhood (Table 7.3). Fula mothers were the least likely to live in their home village at 31%. The majority of mothers (73%) however had daily contact with their birth family. Despite this regular contact, only 55% of the mothers

Table 7.3   Indicators of social support for 396 mothers and 337 fathers, as percent reporting or mean as specified

	Mothers	Fathers
Frequency	396	337
Birth Family		
Lives in home village	48.0	88.1
Family members in village	80.5	96.4
Daily contact with family	72.7	95.0
Friendship and Groups		
Member of a village group	92.4	73.0
Mean # people to call in an emergency	4.3	9.4
Relative Support		
Perceive better than average community support	7.3	11.9
Variables in Parent Support Score		
Strongly agrees – problem solver	40.4	49.9
Strongly agrees – someone to care for family	47.7	47.3
Strongly agrees – spouse would care for children as they do	45.7	73.6
Strongly agrees – family would come in an emergency	54.8	54.0
Mean Parent Support Score	11.1	11.4

strongly agreed a member of their birth family would come in an emergency if they sent for them. While the Fula mothers were also less likely to have daily contact with their birth family (66%) there were no significant differences in their expected ability to call on family members in an emergency. Social support from non-family members in the village was common; 92% of mothers were members of a village group and 59% strongly agreed that they had either an aide to problem solving or someone to care for their family.

The average parent support score for mothers was 11.1 with a range from 5.6 to 12.7. Ethnic differences were not found. Twenty-five percent of the mothers had a parent support score less than 10.39. The mothers who fell into this lowest quartile were also less likely to live in their home village (OR = 0.56, p = 0.015) and less likely to live in a village with family members (OR = 0.37, p = 0.000) compared to mothers in the higher quartiles. These mothers in the lowest quartile were more likely to have a parent SES sum of less than three (OR = 2.74, p = 0.000). The mothers of the lowest quartile did not differ from mothers with higher parent support scores with respect to having attended formal schooling, English literacy, membership in village groups, or their perceptions of relative community support.

A greater proportion of fathers were living in their home villages (88%) compared to mothers (Table 3). Fula fathers were less likely to live in their home village (82%) compared to Mandinka (92%) and Wollof (89%). Nearly all fathers (95%) had daily contact with their birth family but only 54% strongly agreed a member of their birth family would come in an emergency. Fathers of all three ethnic groups were equally as likely to expect to be able to call on their birth family. Fathers were less likely to belong to a village group (73%) than mothers but they were more likely to report support from friendships. Sixty-six percent of fathers strongly agreed they had either or both of an aide to problem solving or someone to care for their family. Unlike the mothers, none of the fathers reported they had no one to call on for help in an emergency. Fathers ranked themselves slightly better than mothers with respect to general relative support from the community and there were no significant differences for fathers by ethnic group.

The average parent support score for fathers was 11.4 with a range from 8.8 to 12.5. No ethnic differences were found. Twenty-five percent of the fathers had a parent support score less than 10.66. The fathers who fell into this lowest quartile were also less likely to live in their home village (OR = 0.55, p = 0.097) and less likely to live in a village with family members (OR = 0.10, p = 0.001) compared to fathers in the higher quartiles. These fathers in the lowest quartile were more likely to have a parent SES sum of less than three (OR = 2.10, p = 0.050). The fathers of the lowest quartile did not differ from fathers with higher parent support scores with respect to having attended formal schooling, English literacy, membership in village groups, or their perceptions of relative community support.

Comparisons between Cases and Controls

Cases and controls were similar according to nearly all of the mothers' sociodemographic characteristics (Table 7.4a). Mothers of cases were more likely to have a possession that could be used to guarantee a loan (p = 0.084). No significant differences were found for compound SES scores. Mothers of cases and controls were equally as likely to have a parent SES sum less than 3. Mothers of cases were less likely to strongly agree that they could expect family members to come for help in case of an emergency (p = 0.013) (Table 7.5a). Mothers of the cases and controls were equally as likely to have parent support scores in the bottom quartile of the distribution.

Cases and controls were also similar according to their fathers' sociodemographic characteristics (Table 7.4b). No significant differences were found for compound SES scores. Fathers of cases and controls were equally as likely to have a parent SES sum less than 3. Reports of family relationships were different between fathers of cases and controls (Table 7.5b). Fathers of cases were more likely to be living in their home village (p = 0.071), to have family members in the same village (p = 0.026) and to have daily contact with their birth family (p = 0.003). Fathers of cases and controls however were equally as likely to strongly agree that family members would come in an emergency. Fathers of controls were more likely to fall in the bottom quartile of the distribution of parent support scores (p = 0.043).

Multivariate Models of Child Mortality

Multinomial logistic regression models were used to predict infant deaths (<1 year of age) and deaths to children aged 1-4 years separately, each compared to controls. Control variables retained in the final models were: parent's age, child's sex, parent's ethnicity, parent's schooling, and membership in the lowest quartile of compound SES. Mother's low SES sum and low parent support score were not associated with

Table 7.4a **Cases and controls with mothers sociodemographic and socio-economic indicators, compared using ANOVA for means and chi-square for proportions**

	Cases	Controls	p-values
Frequency	206	211	
Mother's Age	31.3	30.9	.560
Mother's Education			
Attended Formal School	10.7	7.1	.200
Literate in English	4.4	7.1	.230
Attended Qu'ranic Training	83.5	81.5	.595
Mother's Ethnicity			.389
Fula	20.9	22.7	
Mandinka	41.7	36.5	
Wollof	37.4	39.8	
Mother's Occupation			
Farming only occupation	48.5	48.8	.956
Mother's Children			
Mean No. Children in DSS	5.2	4.9	.347
Mean No. Child deaths in DSS, minus case	0.4	0.4	.388
Variables in Parent SES Sum			
In possession of money	47.1	46.9	.973
Could sell something in emergency	64.9	69.0	.366
Owns something to guarantee a loan	88.8	82.9	.084
Has goods or produce to sell	64.6	58.8	.224
Uses oil for cooking less than once weekly	22.3	21.8	.896
Parent SES Sum < 3	20.9	25.6	.254

either infant or 1-4 deaths in these models. Among the control variables, infant mortality was higher among children whose mothers had attended school (OR = 2.43, p = 0.027). Father's low SES sum was not associated with either infant or 1-4 deaths in these models, using the same control variables. Children of fathers who had the lowest social support scores had lower 1-4 mortality risk (OR = 0.52, p = 0.037) while their infant mortality risk was not significantly different from children of other parents (OR = 0.73, p = 0.310). None of the control variables were significantly associated with either infant or 1-4 deaths in the models using father's characteristics. Additionally, there were no significant interactions between the lowest parent SES sums and parent support scores in either mother's or father's models.

Discussion

There were few factors that proved to be significant predictors of child mortality in this data. This may be partly due to the fact that deaths to children aged less than five are a common experience for parents in the population. Deaths to children were

Table 7.4b   Cases and controls with fathers sociodemographic and socio-economic indicators, compared using ANOVA for means and chi-square for proportions

	Cases	Controls	p-values
Frequency	189	186	
Father's Age	45.3	45.2	.872
Father's Education			
Attended Formal School	3.2	6.5	.138
Literate in English	2.6	6.0	.112
Attended Qu'ranic Training	96.3	92.5	.110
Father's Ethnicity			.379
Fula	20.1	22.0	
Mandinka	41.8	34.4	
Wollof	38.1	43.0	
Father's Occupation			
Farming only occupation	33.9	34.9	.825
Father's Children			
Mean No. Children in DSS	8.2	8.4	.700
Mean No. Child deaths in DSS, minus case	0.7	0.6	.327
Variables in Parent SES Sum			
In possession of money	59.8	64.5	.345
Could sell something in emergency	93.7	94.1	.861
Owns something to guarantee a loan	96.3	95.2	.587
Has goods or produce to sell	74.6	73.1	.743
Uses oil for cooking less than once weekly	24.9	29.0	.363
Parent SES Sum < 3	9.5	10.2	.822

common even among parents of the controls; 1 in 4 mothers and 4 in 10 fathers of controls had child deaths listed in the DSS registry. Parents of cases and controls had similar numbers of deaths to children, among their other children registered in the DSS. Given that death is such a common experience, it may be preferable to consider the determinants of a high numbers of deaths for individual parents. By classifying parents according to individual children, we may have allowed parents of cases and parents of controls to be too similar with respect to their overall experiences with child mortality.

As an alternative to the models predicting mortality for individual children using parent's characteristics, we considered modelling 3 or more deaths as an outcome variable for parents. Having 3 or more deaths was only associated with parent's age in multivariate models. This effect may have been picking up time rather than the true effect of age. Parents in the full DSS had a maximum of 4 deaths listed for mothers and 7 deaths listed for fathers. DSS data may not be the best data to consider the determinants of the death clustering for parents because of this tight range of the number of deaths. Cross-sectional surveys listing children ever born for adults in this population show more evidence of clustering with as many as 9 deaths for women

Table 7.5a Cases and controls with mothers social networks, compared using ANOVA for means and chi-square for proportions, with p values reported

	Cases	Controls	p-values
Frequency	206	211	
Birth Family			
Lives in home village	48.5	46.0	.599
Family members in village	80.6	79.9	.862
Daily contact with family	73.3	71.6	.692
Friendship and Groups			
Member of a village group	91.7	92.9	.661
Mean # people to call in an emergency	4.5	4.0	.110
Relative Support			
Perceive better than average community support	6.3	9.5	.231
Variables in Parent Support Score			
Strongly agrees – problem solver	42.7	37.4	.271
Strongly agrees – someone to care for family	44.2	49.8	.253
Strongly agrees – spouse would care for children as they do	42.2	49.8	.123
Strongly agrees – family would come in an emergency	49.0	61.1	.013
Parent Support Score < 10.39	26.2	23.7	.553

(aged 15-54) and 14 deaths for men (aged 18+) among their full set of children (Ratcliffe, unpublished). Due to migration of children and time living away from parents, the DSS registry data may not be complete enough to pick up death clustering for parents.

The fact that parents were allowed to repeat when siblings were included among the sets of both cases and controls may limit the ability to detect effects of indirect factors on child mortality. While it was deemed unacceptable to let siblings repeat among the cases and not among the controls, the inclusion of siblings among the set of cases and controls meant that 9 women and 21 men were matched to one of each. This repetition would have dampened any associations between parents' characteristics and child mortality.

Associations between child mortality and socio-economic status of parents may be weak in this population since, in this subsistence economy, the range of SES represents only degrees of poverty. While household production of groundnuts ranged from 1-200 sacks, the nature of the market with a single payment some months after harvest means that even the better off families find themselves surviving on very little for a large part of the year. The season of highest child deaths, corresponding to malaria season, also falls just before and after harvest when resources are lowest. As an indication of the scarcity of money during this time of year, we note that less than two-thirds of the fathers and half of the mothers had money in their possession at

Table 7.5b Cases and controls with fathers social networks, compared using ANOVA for means and chi-square for proportions, with p values reported

	Cases	Controls	p-values
Frequency	189	186	
Birth Family			
Lives in home village	91.0	84.9	.071
Family members in village	98.4	94.0	.026
Daily contact with family	98.4	91.9	.003
Friendship and Groups			
Member of a village group	71.4	77.4	.184
Mean # people to call in an emergency	9.4	9.5	.936
Relative Support			
Perceive better than average community support	11.6	11.8	.955
Variables in Parent Support Score			
Strongly agrees – problem solver	52.9	47.3	.278
Strongly agrees – someone to care for family	49.5	45.2	.404
Strongly agrees – spouse would care for children as they do	76.2	72.0	.359
Strongly agrees – family would come in an emergency	57.1	55.4	.730
Parent Support Score < 10.66	19.6	28.5	.043

interview. The parent's personal SES sum however was created to pick up the individuals with available resources, such as money of their own or access to money through a loan. The fact that this indicator was not associated with child mortality among children of mothers or fathers may support the adage that money does not buy health.

In the multivariate models, a significant association was found between mother's schooling and child deaths. Children whose mothers had attended any formal school were more likely to have died as infants. While this finding goes against many public health messages, similar findings have been found in other subsistence economies in Africa (e.g. Van den Broeck, 1996). Mothers who had been to school may at first be thought to be at a disadvantage because their education sets them apart from others in their community but there was no difference between the social support scores of mothers with respect to education. While 8% of the women had attended school, levels of schooling completed were low and only 7 women had completed primary school. The mothers' schooling effect on mortality was not found for children over age 1.

It is also interesting to note that education among the fathers was lower than among the mothers, contrary to known gender differences in education for the study area (Ratcliffe et al., 2002). The age of the sample is likely to have shaped this finding. Education levels are highest among the youngest age groups. This sample includes parents who will tend to be older on average than the general population of adults.

Additionally, fathers in this sample were older than mothers and this may explain their lower education. Age and education were both included as control variables in the multivariate models.

Despite the fact that the effect of parental factors on child mortality may have been dampened for the reasons explained above, a significant effect was found for father's social support as a predictor of 1-4 mortality even controlling for other factors in the multivariate models. The effect however was not in the direction expected. Children of fathers with better social support were at higher risk of 1-4 mortality. We hypothesised that social support would allow parents better access to resources for health maintenance and promotion. Social support is based on a set of reciprocal relationships and therefore may cost individuals as well as benefit them. The finding that better social support is associated with 1-4 mortality may at first seem due to a strain on men's resources rather than improved access to others' resources. The finding that SES is not associated with 1-4 mortality however makes it seem unlikely that availability of economic resources is an important factor for child survival in this population. The explanation of the increased risk of mortality among children whose fathers have better social support must lie elsewhere.

Men's social support may be more closely linked to family dynamics of health decision-making. For fathers, living in one's home village and having better contact with family members was also associated with higher child mortality. The context of family networks may deserve particular attention. Men with more social support may be less inclined or less able to act in times of need for child health. Alternatively, men's social networks may hamper health-promoting behaviours of other members of the family. The roles of various family members in decision-making around child health could provide possible pathways for health promotion at the household level.

No association was found between father's parent support score and infant mortality. It is difficult to imagine that members of men's social networks or men themselves would play different roles in decision-making around infant health compared to 1-4 health but it is likely that the different set of factors that influence infant health and disease may make these roles less important to survival. No associations were found for mother's social support and infant or 1-4 mortality. Gender differences in social networks as well as different roles for men and women in the decision-making processes around health for children under age five may be likely to explain such effects.

Social support in this population was high and family ties were strong for both parents. Residence in one's home village was found to be important not only for support from family but also for other forms of practical support included in the parent support score. Fathers' better overall support compared to mothers may be due to migration at marriage in the population with women moving to their husbands' home. Even with this migration, women often have means of social support available. The ability to call on family members in times of emergency did not correspond to residence in the home village so even when family were living far away, they were still considered a resource for support in times of need. Women also had access to new social networks in the husbands' communities and membership in village groups was more common among women than men.

Despite the fact that for nearly everyone in the population SES is low and social support is strong, there are still important features of the poor families without strong social ties that distinguish them from others. Individuals with poor social support were also more likely to suffer from low SES. Given these individuals had weak ties to family and limited means beyond farming to improve their lot, they are likely to remain disadvantaged. The daily struggles of these individuals and the impact on the well-being of their families should not be discounted, even though no association was found for child mortality.

It is striking that no benefit of social support was found for child survival in this population. Social capital has been promoted as a mediator of the relationship between socio-economic status and health status at the community level mostly in developed countries (Kawachi et al., 1997). If we are able to liken the family to the community, this theory may be applied to consider that families with strong social support may be better able to overcome scarce resources and poor SES to provide for health of family members. Even where families are concerned about ill health however, social networks may not benefit health if the other members of these networks have little access to, experience with, and knowledge of health promoting behaviours.

With respect to health equity for child survival, family socio-economic status and social support cannot be considered as stratifiers for this population. Deaths to children under age five are common, and the ranges of both SES and social support within the population are tight. No support was found for the hypothesis that SES and social support mattered through resource availability for child health promotion. Better social support did not benefit child survival. Since socio-economic status is not associated with mortality and given the fact that father's better social support increased risk for children aged 1-4, it seems that the pathway connecting social support and child mortality is likely to be through health-seeking behaviour and decision-making at the family level.

References

Kawachi, I. and Kennedy, B.P. (1997). 'Health and social cohesion: why care about income inequality?' *BMJ* 314, 1037-1040.

Kawachi, I., Colditz, G.A., Ascherio, A., Rimm, E.B., Giovannucci, E. Stampfer, M.J. et al. (1996) 'A prospective study of social networks in relation to total mortality and cardiovascular disease in men in the US'. *Journal of Epidemiology and Comm Health* 50, 245-251.

Mosley, W.H. and L.C. Chen (1984). 'An Analytical Framework for the Study of Child Survival in Developing Countries'. *Population and Development Review.* 10 Supp(84): 25-45.

Oppong, C. (1992). 'Traditional Family Systems in Rural Settings in Africa'. In E. Berquo and P. Xenos (eds). *Family Systems and Cultural Change,* Clarendon Press: Oxford.

Ratcliffe, A.A. (2000). *Men's Fertility and Marriages: Male Reproductive Strategies in Rural Gambia.* Doctoral Dissertation, Harvard University: Boston, USA.

Ratcliffe, A.A., A.G. Hill, P. Gomez, and G. Walraven (2001). 'Farafenni, The Gambia'. In INDEPTH Network. Health demographics in developing countries. Volume 1. Population, health, and survival at INDEPTH sites. International Development Research Centre, Ottawa, Canada.

Van den Broeck, J., R. Eeckels, G. Massa (1996). 'Maternal Determinants of Child Survival in a Rural African Community'. *International Journal of Epidemiology* 25(5): 998-1004.

Chapter 8

Health and Health Care: Equity Aspects in FilaBavi, Vietnam

Nguyen Duy Khe, Pham Huy Dung, Ho Dang Phuc,
Hoang Van Minh, Nguyen Xuan Thanh, Bo Eriksson,
Vinod Diwan and Nguyen Thi Kim Chuc

Summary

This study describes health status, health seeking behavior and expenditure on health of household members living in one rural district in northern Vietnam. Three data sets from three cross-sectional and longitudinal studies were used. Data on morbidity, mortality, health care utilization and health care expenditure were collected by follow-up surveys, while information on household economic status was collected by baseline surveys. Morbidity, mortality patterns as well as utilization of and expenditures on health care of persons in different economic groups were assessed.

This chapter highlights several important issues currently of concern to health policy makers and planners related to equity in health and health care. Males and both males and females of working age in lower economic groups have higher mortality rates than those in higher economic groups. Using self-reported morbidity to measure health status, the picture of health is different and is of more complex patterns: women have higher morbidity frequency than men and the poor reported to be sick more often than the better-off. However the study points out that people from richer quintiles used hospital and private practitioners more often than the poor did. This study also reveals similar findings to that of an earlier study carried-out in the same setting using a cross-sectional design; self-treatment was a common practice and no-treatment was reported more often in poorer quintiles. Our results also showed that payments for health services were substantial for households. The proportion of expenditure for health of households in the poorest quintile was significantly higher than that of households in the least poor quintile.

Background

In 1986, the Vietnamese government initiated a wide-ranging economic reform program known as DoiMoi. At the time, the resulting economic development has had many positive impacts on health due to the improvement of living standards and education. In general, in urban as well as in rural areas the number of poor households (with income not enough to provide meals of 2,100 calories/person/day) decreased from 55% in 1989 to 19.9% in 1993.[1] During the period from 1990 to 1996, GNP per capita reached 290 USD, with average annual growth rate of 6.2 %.[2]

However, Vietnam is still a poor country. According to the 2000 World Health Report of the WHO, Vietnam is the 116th country for DALE among 130 countries, and stays at 50th place in health financing ranking of 187 countries.[3] In 2000, about 20% of the population lived in very poor conditions; most of them living in mountainous and remote areas.[4]

Among the most important measures of health reform in the health sector, the introduction of fees for health services at higher level of public health facilities (e.g. hospital) and health insurance policy have made the health system more responsive while health services become less accessible for poor people. More than 30% of people in poor areas did not seek care when they were last sick. In fact, it was estimated that there are about 28 million poor people (out of a 76 million's population) who can not pay the user fees and do not belong to the group which has the fees waived.[5]

This study is an attempt to use available secondary data collected from the epidemiological field laboratory, called FilaBavi to analyze the health equity situation among persons with different economic status.

The specific objectives are to:

1. Assess and compare morbidity and mortality pattern of persons with different economic status.
2. Assess and compare health seeking behavior and expenditure on health of persons with different economic status.
3. Contribute to a discussion concerning the relationship between health, health equity and economic reform and to provide input to the policy formulation processes.

Methods

Study Site and Study Sample

This study used the data collected within the Epidemiological Field Laboratory of Bavi (FilaBavi), which is located in Bavi district of Ha Tay province in the North of Vietnam. The site is situated 60 km to the northwest of Hanoi, the capital city. The

1   National committee for population and family planning of Vietnam 1998. Population Studies and Information 1998.
2   UNICEF 1999. Unicef Annual Report 1999.
3   World Health Organization, 2000. The World health's report 2000. WHO, Geneva.
4   Pham Manh Hung, I. Harry Minas, Yuanli Liu, Goran Dahgren and William Hsiao. 'Efficient-Equity-oriented strategies for health – International perspective – focus on Vietnam'. CIMH 2000.
5   Do Nguyen Phuong, 'Issues of equity and effectiveness in health care in Vietnam, Efficient-Equity-oriented strategies for health – International perspective – focus on Vietnam'. CIMH 2000.

district covers an area of 410 km² including lowlands, highlands and mountainous areas. It consists of 32 communes with a population of approximately 235,000 people belonging to three major ethnic groups: Kinh (91 %), Muong (8 %), Dao (Man) (<1%) and some families of Tay, Hoa (Chinese) and Khmer tribe groups. In 1996, the annual income per capita of Bavi was 290 kg rice (about 600 000 VND ~ 48 USD).

In frame of FilaBavi, a random sampling of units, proportional to population size was performed; including 69 clusters (11,089 households) with a population about 50,000 inhabitants (approximate 20% of the total district inhabitants). The sampling units (clusters) cover all different geographical regions in the district.

The data collecting process in FilaBavi is performed by 43 surveyors divided into 6 groups led by field supervisors. The surveyors, visiting households, conducted interviews and filled in suitable surveillance forms. 'Spot checks' by field supervisors were conducted to witness the interviews on approximately a 5 percent sample of the home visits per cycle. Feedback was given to the interviewer by the field supervisor.

The general socio-economic information of the households and individuals were collected in the baseline survey and recencuses (conducted every two years) with a main questionnaire. Follow-up surveillance has been carried out quarterly concerning all events, changes that have happened in the family during the previous three months including death, birth, pregnancy, marriage, migration and illness.

Data Analysis

In this study the data used were taken from three sources within FilaBavi:

- Data related to the economic situation of households were extracted from Baseline Survey (January to April 1999) and First Recensus Survey (from April to June 2001).
- Information on mortality was collected by 13 follow-up surveys (from May 1999 to August 2002). Data related to morbidity was collected in 10 follow-up surveys (with the gap of 3 surveys in the period April-December 2001).
- Data from a sub-study on health care expenditure from July 2001 to June 2002. In this sub-study, 621 households were randomly selected from all clusters of FilaBavi and the household's heads were interviewed monthly giving the information on health care utilization and health expenditure.

To classify households into different economic levels, a parameter called 'wealth index' was employed. The wealth index has been calculated by using method of principal component analysis based on all variables related with the economic situation of households such as: housing conditions, household assets, sanitary conditions, land area, incomes and expenditures. All categorical household economic variables were recoded into binary sub-variables before the computation. Missing values of every variable were replaced by mean values. The first principal component was employed as a wealth index.[6]

---

6   Jolliffe, I.T. *Principal Component Analysis*. New York 1986.

Thus, the wealth index is a linear combination of all household economic variables. In this combination, the coefficients of the variables were chosen so that the wealth index could best represent the household economic status. In other words, the wealth index could enable the differentiation of households in terms of economic levels. Five quintiles of the index represent the economic status of households as well as household's members.

The WHO standard population[7] was used to calculate age standardized rates, which were taken to compare mortality, morbidity and health service utilization between wealth quintiles.

The data analysis was performed using statistical packets STATA 7.0 and SPSS 10.0.

Definitions

A case of illness was defined as follows:

- The illness episode started earlier or within a 4-week period prior to the interview, recovery was complete at the time of the interview and illness time covered a noticeable part of the period.
- The illness episode started before or within a 4-week period, and recovery was not complete at the time of the interview.
- A person who fell sick more than one time during the 4-week period, due to the same or different diseases would be counted as two or more cases.

Different *types of health care* were defined as follows:

- *Self-treatment*: patient who treated themselves using medicines available at home or bought from drug sellers without any medical examination. Patients who took medicament (modern or traditional) following the advice given by a family member, a relative, a friend, a neighbor or any person without medical background.
- *Traditional healer*: patients who sought health care from traditional healers.
- *Private health care*: patients who sought health care from private clinics. Also includes cases who went to public health workers practicing privately after regular work or to retired health workers practicing at home.
- *Community health station (CHS)*: patients who visited CHS, including patients who had referral letters to higher levels.
- *District health center*: includes different kinds of hospitals as well as other public health facilities at the district level (tuberculosis clinic, dermatology clinic, etc).
- *Provincial/central hospital*: includes various hospitals at provincial and central level.
- *Others*: any health care service other than the above mentioned is counted in this category.

---

7   WHO. Age Standardization of Rates: A New WHO Standard. GPE Discussion Paper Series: No. 31. 2000.

Results

Mortality Pattern in FilaBavi

During three and a half years of surveillance in FilaBavi there were 860 deaths registered from a total of 175,380 person-years of follow-up. The population distribution and age specific mortality rates are shown in Table 8.1. The crude mortality rate is 4.90 per 1000 person-years. The rates for males and females are 5.42 and 4.43 respectively.

To study the impact of economic condition on mortality rates, the age standardized mortality rates are calculated for 5 economic groups of people classified by the wealth index. The results illustrated by Figure 8.1 show that the age standardized mortality rates of whole population and of females are not significantly different across the wealth quintile groups. However, the rates of males from the poorest quintile are much higher than the ones of males from the richest and IV quintile. The ratios of the rate from first quintile to the two last quintiles are 1.72 and 1.67 with 95% confidence intervals (1.34-2.21) and (1.30-2.16).

When age was regrouped into: children under 15 years, people of working age (15 to 59 years old) and elderly (60 years old and over), the results show that for

Table 8.1   Age specific mortality rates

Age (years)	Person-years Total	Male	Female	Death Total	Male	Female	Rate (per 1000 PY) Total	Male	Female
<1	2653	1376	1277	51	32	19	**19.23**	**23.26**	**14.88**
1-4	11191	5806	5385	15	10	5	**1.34**	**1.72**	**0.93**
5-9	17727	9220	8507	12	8	4	**0.68**	**0.87**	**0.47**
10-14	21232	10719	10513	7	4	3	**0.33**	**0.37**	**0.29**
15-19	21074	10843	10231	11	5	6	**0.52**	**0.46**	**0.59**
20-24	13736	6604	7131	10	4	6	**0.73**	**0.61**	**0.84**
25-29	12696	6144	6552	14	10	4	**1.10**	**1.63**	**0.61**
30-34	11945	5732	6214	9	5	4	**0.75**	**0.87**	**0.64**
35-39	13731	6680	7051	19	17	2	**1.38**	**2.54**	**0.28**
40-44	12096	5732	6364	33	24	9	**2.73**	**4.19**	**1.41**
45-49	8139	3799	4340	30	22	8	**3.69**	**5.79**	**1.84**
50-54	5592	2608	2984	27	17	10	**4.83**	**6.52**	**3.35**
55-59	4568	2140	2428	27	19	8	**5.91**	**8.88**	**3.29**
60-64	4907	2115	2792	46	28	18	**9.37**	**13.24**	**6.45**
65-69	4665	1813	2852	79	50	29	**16.93**	**27.58**	**10.17**
70-74	3967	1500	2467	110	66	44	**27.73**	**44.01**	**17.84**
75-79	2829	905	1924	113	53	60	**39.94**	**58.60**	**31.18**
80-84	1607	411	1196	92	38	54	**57.24**	**92.56**	**45.12**
>84	1024	230	793	155	45	110	**151.37**	**195.23**	**138.63**
Total	**175380**	**84377**	**91003**	**860**	**457**	**403**	**4.90**	**5.42**	**4.43**

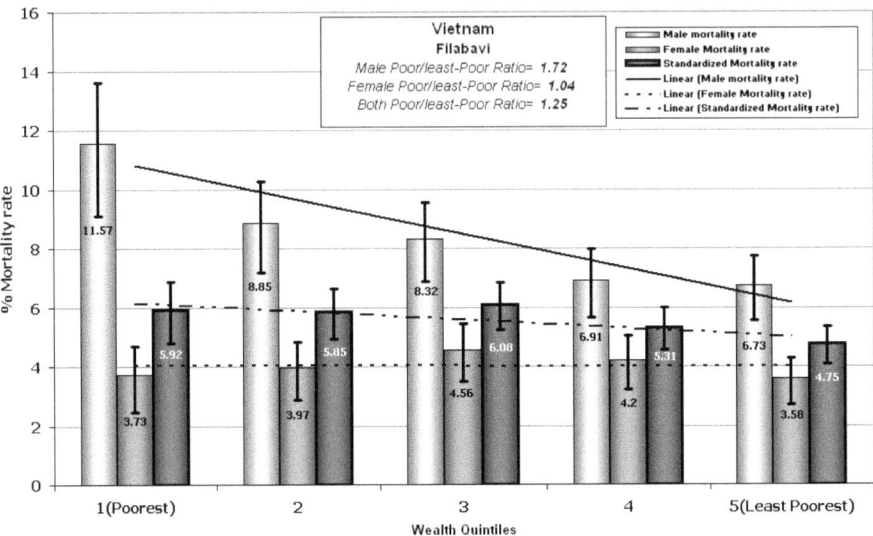

Figure 8.1   Age standardized mortality rates with 95% CI across wealth quintiles

the working age group there are mortality differentials between wealth quintiles as a whole and among men and women of different wealth quintiles (Table 8.2).

The rate ratios of the poorest wealth quintile to the least poor quintile is 3.54 (3.63 and 3.26 for males and females respectively).

Morbidity pattern in FilaBavi

During 10 follow-up surveys, there were 192,479 cases of illness reported with 38,518 person-years (each person contributed at most one month of follow-up time at every survey). The general estimated number of sickness episodes per person was 5.0 times per year.

To study the influence of economic condition on health, the number of illness episode/person/years were calculated for 5 groups of wealth indices. To avoid age confounding (as commonly known, elderly people and children are the persons most suffering from sicknesses), the age standardizing method was used in calculating the morbidity frequency.

The result shows significant morbidity differential between wealth quintiles for whole population (Figure 8.2). The poorest people fell ill 6.50 times per year in average. The morbidity frequency decreased across economic groups and reduced to 4.42 for the group of people with the highest economical status.

When considering males and females separately or individually, there are morbidity differentials. In general, women reported to be sick more often than men; and men

Table 8.2   Age standardized mortality rates of children, persons of working age and elderly people

			Wealth Index Quintiles					Total
			I	II	III	IV	V	
Children under 15 years old	Male	Rate	3.66	2.37	2.20	2.85	1.94	2.48
		95% CI	2.1-6.3	1.4-4.1	1.3-3.7	1.8-4.6	1.1-3.5	2.0-3.2
	Female	Rate	2.09	0.76	2.46	2.25	1.84	1.53
		95% CI	1.0-4.4	0.3-2.0	1.5-4.2	1.3-4.0	1.0-3.4	1.2-2.1
	Both Genders	Rate	2.88	1.57	2.33	2.55	1.89	2.02
		95% CI	1.9-4.5	1.0-2.5	1.6-3.4	1.8-3.7	1.2-2.9	1.7-2.4
Working people (15-59 years old)	Male	Rate	7.59	4.89	4.26	1.78	2.09	2.99
		95% CI	5.5-10.5	3.6-6.7	3.2-5.7	1.2-2.7	1.5-3.0	2.5-3.5
	Female	Rate	2.19	1.56	1.81	0.97	0.67	1.25
		95% CI	1.3-3.7	0.9-2.7	1.2-2.8	0.6-1.7	0.4-1.3	1.0-1.6
	Both Genders	Rate	4.89	3.23	3.04	1.38	1.38	2.07
		95% CI	3.8-6.4	2.5-4.2	2.4-3.9	1.0-1.9	1.0-1.9	1.8-2.4
Elderly Persons (60 years old and over)	Male	Rate	49.65	43.70	42.91	42.52	41.46	43.43
		95% CI	35.5-69.5	32.3-59.2	33.3-55.3	34.1-53.1	33.5-51.3	38.8-48.6
	Female	Rate	15.36	23.56	23.44	25.29	22.57	22.20
		95% CI	11.0-21.3	17.4-31.9	17.8-30.8	19.9-32.1	17.75-28.7	19.7-25.0
	Both Genders	Rate	32.50	33.63	33.18	33.91	32.01	29.31
		95% CI	26.6-39.6	27.4-41.3	27.7-39.7	28.9-39.7	27.4-37.4	27.0-31.9

and women in lower economic groups get sick more often than their counterparts in higher economic groups.

Health care utilization and health care expenditure

Table 8.3 shows that self-treatment was common practice and accounted for almost 50% of the health care acts. The rich used private health services more often than the poor (32,42% versus 30,01%) did. Non-treated cases accounted for 8.53% in the group of poorest people. The rate significantly decreased in higher economic groups and was only 4.33% in the richest group.

Similarly, better-off people more regularly use the district health center and the provincial hospital. In the first three groups, the utilization frequencies of these types of health care were significantly less.

Table 8.4 shows utilization of health services by men and women. Men used hospitals significantly more often and treated themselves less than women did.

Table 8.5 shows the average payment for different health services. In general, the poor paid as much as the rich, except in the case of private health care service, the average cost spent for one time of utilization is not different across economic groups.

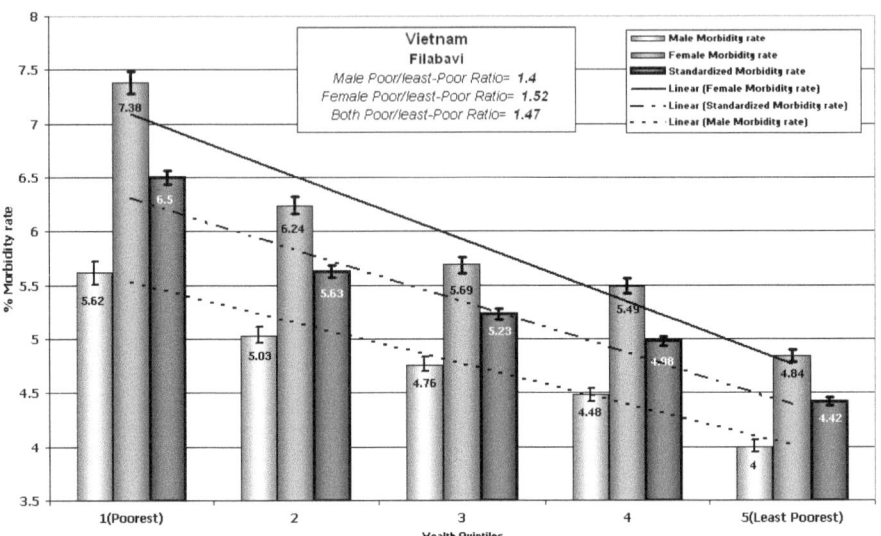

Figure 8.2 Age standardized morbidity rates with 95% confidence intervals

Table 8.3 Health service utilization by wealth index quintiles

		Wealth Index Quintiles					Total
		I	II	III	IV	V	
No Treatment	%	8.53	6.10	5.37	4.80	4.33	5.46
	95% CI	8.2-8.9	5.8-6.4	5.1-5.6	4.6-5.0	4.1-4.5	5.36-5.57
Commune Health Station	%	4.86	4.64	4.77	4.57	4.45	4.65
	95% CI	4.6-5.1	4.4-4.9	4.6-5.0	4.4-4.8	4.2-4.7	4.55-4.75
District Health Center	%	2.92	3.02	3.17	4.07	4.38	3.43
	95% CI	2.7-3.1	2.8-3.2	3.0-3.4	3.9-4.3	4.2-4.6	3.35-3.51
Provincial Hospitals	%	0.90	1.15	1.29	1.57	2.30	1.45
	95% CI	0.8-1.0	1.1-1.3	1.2-1.4	1.5-1.7	2.1-2.5	1.39-1.5
Traditional Healer	%	1.86	1.70	1.73	1.83	2.12	1.88
	95% CI	1.7-2.0	1.6-1.8	1.6-1.9	1.7-1.96	2.0-2.3	1.82-1.94
Self-Treatment	%	50.79	49.52	49.20	50.08	49.59	49.92
	95% CI	50.0-51.6	48.8-50.3	48.5-49.9	49.4-50.8	48.9-50.3	49.6-50.24
Private Facilities	%	30.01	33.68	34.21	32.78	32.42	32.94
	95% CI	29.4-30.6	33.1-34.3	33.6-34.8	32.2-33.3	31.9-31.0	32.7-33.2

Table 8.4   Health services utilization by sexes and wealth index quintiles

			Wealth Index Quintiles					Total
			I	II	III	IV	V	
Without Treatment	Male	%	9.16	6.28	5.55	4.63	4.22	5.50
		95% CI	8.6-9.8	5.9-6.7	5.2-5.9	4.3-5.0	3.9-4.5	5.3-5.7
	Female	%	7.89	5.91	5.18	4.96	4.43	5.53
		95% CI	7.5-8.3	5.6-6.3	4.9-5.5	4.7-5.3	4.2-4.7	5.4-5.7
Commune Health Station	Male	%	4.75	4.80	4.75	4.57	4.44	4.61
		95% CI	4.4-5.2	4.5-5.2	4.4-5.1	4.3-4.9	4.2-4.8	4.5-4.8
	Female	%	4.96	4.47	4.79	4.56	4.46	4.65
		95% CI	4.7-5.3	4.2-4.8	4.5-5.1	4.3-4.8	4.2-4.7	4.5-4.8
District Health Center	Male	%	3.43	3.74	3.93	4.92	5.17	4.43
		95% CI	3.1-3.8	3.4-4.1	3.6-4.2	4.6-5.3	4.8-5.5	4.3-4.6
	Female	%	2.41	2.29	2.42	3.22	3.59	2.81
		95% CI	2.2-2.6	2.1-2.5	2.2-2.6	3.0-3.5	3.4-3.8	2.7-2.9
Provincial Hospitals	Male	%	1.07	1.36	1.58	1.70	2.49	1.72
		95% CI	0.9-1.3	1.2-1.6	1.4-1.8	1.5-1.9	2.3-2.7	1.6-1.8
	Female	%	0.72	0.95	0.99	1.44	2.11	1.28
		95% CI	0.6-0.9	0.8-1.1	0.9-1.1	1.3-1.6	1.9-2.3	1.2-1.4
Traditional Healer	Male	%	1.94	1.65	1.75	1.88	2.09	1.88
		95% CI	1.7-2.272	1.5-1.9	1.6-2.0	1.70- 2.1	1.9-2.3	1.8-2.0
	Female	%	1.79	1.75	1.71	1.78	2.15	1.85
		95% CI	1.6-2.0	1.6-1.9	1.5-1.9	1.6-2.0	2.0-2.3	1.8-1.9
Self-Treatment	Male	%	49.27	47.78	47.62	48.77	47.93	47.89
		95% CI	48.0-50.6	46.7-48.9	46.6-48.7	47.8-49.8	46.9-48.9	47.4-48.4
	Female	%	52.30	51.26	50.79	51.38	51.26	51.11
		95% CI	51.3-53.3	50.3-52.2	49.9-51.7	50.5-52.3	50.3-52.2	50.7-51.5
Private Facilities	Male	%	30.20	34.18	34.52	33.25	33.23	33.67
		95% CI	29.2-31.3	33.3-35.1	33.7-35.4	32.4-34.1	32.4-34.1	33.3-34.1
	Female	%	29.81	33.19	33.89	32.30	31.61	32.51
		95% CI	29.1-30.6	32.4-34.0	33.1-34.7	31.6-33.0	30.9-32.3	32.2-32.9

The expenditure for one time using private health care service in the three lower level groups is significantly smaller than the one in higher two groups.

Hospitalization is the most expensive service. One single visit to provincial hospital and district hospital cost average 250,000 VND and 90,000 VND respectively. Traditional healers are also not cheap compared to modern private practitioners, communal health centers and self-medication. One visit to traditional healers cost about 75,000 VND.

The variations between economic groups in health expenditures as percentage of total expenditures are shown in Table 8.6. In average, the poorest households

Table 8.5   Mean value of the expenditure for one service utilizing

| | | Wealth Index Quintiles | | | | | Total |
		I	II	III	IV	V	
Commune	1000 VND	18	20	14	23	24	21
H-Station	95% CI	11-25	15-25	10-18	18-28	18-29	18-23
District	1000 VND	54	97	98	92	97	89
H-Center	95% CI	38-70	69-124	68-128	66-117	67-127	77-102
Provincial	1000 VND	317	262	173	330	215	242
Hospitals	95% CI	0-710	127-398	95-250	186-474	122-308	185-299
Traditional	1000 VND	64	65	76	86	86	79
Healers	95% CI	44-84	43-87	54-98	51-121	60-113	65-94
Self-	1000 VND	11	9	15	13	15	13
Treatment	95% CI	9-14	7-10.7	11.7-18	10-17	10.9-20	11-14
Private	1000 VND	18	22	21	27	33	25
Facilities	95% CI	16-20	19-24.6	18-23	24.7-30	25-40	23-27

spent 8.4% of total expenditures for their health care needs. Meanwhile, the richest households spent only 6.7% of their total expenditures for health care.

Discussion

This study describes health status as well as health seeking behavior and expenditure on health of household members living in one rural district in northern Vietnam. In the study, both cross-sectional and longitudinal designs were used. Data on morbidity, mortality, health care utilization and health care expenditure were collected by follow-up surveys, while information on household economic status was collected by baseline surveys using interviews. This is the first study in Vietnam assessing morbidity and mortality patterns as well as utilization of and expenditures on health care of persons in different economic status using data collected by longitudinal surveys. Our paper highlights several important issues currently of concerns to health policy makers and planners related to equity in health and health care.

Table 8.6   Health care expenditure shares in household's expenditure

| | Wealth Index Quintiles | | | | | Total |
	I	II	III	IV	V	
Mean of proportions	8.4%	7.2%	6.6%	6.7%	6.7%	7.1
95% CI	7.77-9.08%	6.58-7.74%	6.03-7.11%	6.14-7.25%	6.04-7.37%	6.88-7.42

A relation between socio-economic status (SES) and health has been widely documented in epidemiological studies.[8] Socio-economic differentials in morbidity and mortality are well documented in research in most European countries and North America. Regardless of which SES indicator or the health outcome measure was used, the link between SES and health was found. Lower socio-economic status is associated with poorer health.[9] [10] [11] [12] [13] There are however relatively few studies on health differentials, especially mortality inequality in low-income countries.[14]

Health Status Differentials

Each member of a community has the equal right to good heath. In our study, morbidity – episode of illness within four week's period and mortality were used as health status indicators. With regard to mortality, the results indicated that males and people of working age in lower economic groups have higher mortality rates than those in higher economic groups. There are several explanations for the differences in mortality rates. One explanation for higher mortality rates of men in lower economic groups compared with men in higher economic groups is the richer have more opportunities to use better quality and higher levels of care services. Also there are better living conditions of people in higher economic groups; for example, better housing with better hygiene and less crowding would lead to lower mortality from communicable disease with fewer preventable deaths.[15] The other reason may that men and working people of both sexes in lower economic groups are more likely to be exposed to occupational hazards compared with people in higher economic groups.

8   Duncan, B.B., Rumen, D., Zelmanovicz, A., Mengue, S.S., Dos Santos, S. and Dalmaz, A. 'Social inequity in mortality in Sao Paulo State, Brazil'. International Journal of Epidemiology. 1995; 24(2):359-356.

9   Karin, H., Humphries and Eddy van Doorlear. 'Income-related health inequality in Canada'. Soc Sci Med 2000; 50:663-671.

10 Kaplan, C.A. and Kei, J.E. Socioeconomic factors and cardiovascular disease: a review of the literature. Circulation 1993; 88 (4 Pt 1):1973-98.

11 Lynch, J.W., Kaplan, G.A., Cohen, R.D., Tuomilehto, J. and Salonen, J.T. 'Do cardiovascular risk factors explain the relation between socioeconomic status, risk of all-cause mortality, cardiovascular mortality, and acute myocardial infarction?' Am J Epidemiol 1996; 144(10):934-42.

12 Wilkinson, R.G. 'Socioeconomic determinants of health. Health inequalities: relative or absolute material standards?' BMJ 1997; 16:93-112.

13 Van Doorslaer, E., Wagstaff, A., Bleichrodt, H., Calonge, S., Gerdtham, U.G., Gerfin, M., Geurtus, J., Gross, L., Hakkinen, U., Leu, R.T., O'Donnell, O., Propper, C., Puffer, F., Rodriguez, M., Sundberg, G. and Winkelhake, O. 'Income-related inequalities in health: some international comparisons'. Journal of Health Economics 1997; 16:93-112.

14 Makinen, M., Waters, H., Hasan, Z.M. and Zaman, S.S. 'Measurement of socioeconomic status for child health research: comparative results from Bangladesh and Pakistan'. Soc Sci Med 1994; 38(9):1287-97.

15 Jun Gao, Juncheng Qian, Shenglan Tang, Bo Eriksson and Erik Blas. 'Health equity in transition from planned to market economy in China'. Health Policy and Planning 2002; 17:20-29.

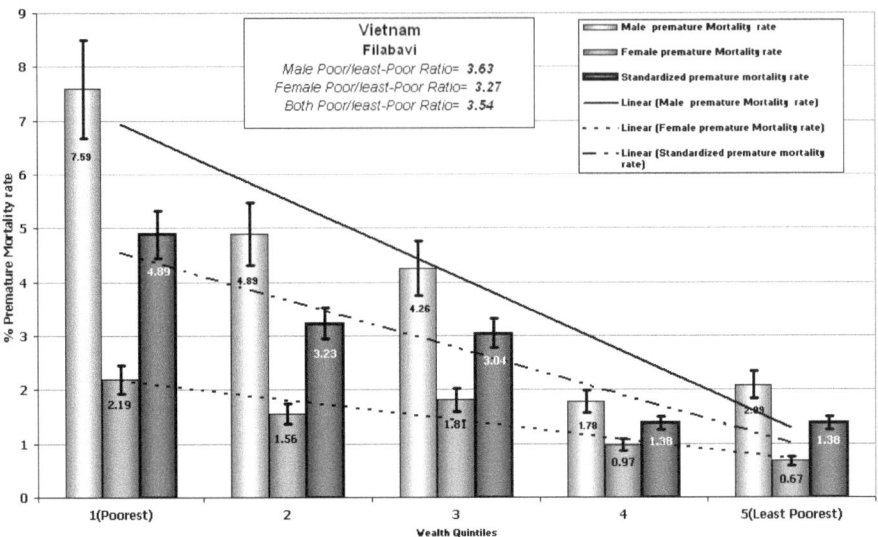

Figure 8.3   Standardized premature mortality rates (for 15-59 years old people) by wealth quintiles

In terms of morbidity, the number of illness episode/person/year was estimated and to avoid age confounding, the age standardizing method was used in calculating the morbidity frequency. Unlike mortality, which is rather blunt indicator of health status, using self-reported morbidity to measure health status, the picture of health is different and is of more complex patterns. Namely, women have higher morbidity frequency than men and the poor reported to be sick more often than the better-off (annual illness episodes were highest among the poorest quintile compared to four remaining quintiles – the ratio between poorest and richest quintile is 1.47). The pattern of morbidity frequency between males and females and between the poor and the better off was similar to studies from other countries.[16] [17] In general women have more complaints on health than men. That finding can partly be explained by the difference between men and women in terms of biological characteristics. Furthermore, women often suffer from the consequences of the reproductive health. However, this finding does not necessarily indicate that women are at greater risk of illness than men. It may reflect the fact that women are more aware of their illness and better at reporting them. In 1993, a national-wide cross-sectional survey called Vietnam Living Standard Survey (VNLSS 1993) was carried out by General Statistics

---

16   World Bank. The health status of adults in developing countries. Washington DC World Bank 2000.
17   Shanlian Hu, Xiaoming Cheng, Xianggang Gong and Dongmei Ying. 'Financing and delivery of health care for the rural population in China'. In: Hung, P.M., Minas, I.H., Liu, Y., Dahlgren, G., Hsiao, William C. (eds). Efficient, Equity-oriented strategies for health: International perspective-focus on Vietnam. CIMH Melbourne 2000.

Office in Vietnam and the survey showed differences in self-reported morbidity between poorest and richest expenditure quintiles but the difference was not significant.[18] The morbidity frequency in our study was also higher than that of VNLSS 1993 (5,24 versus 3,3). Several reasons may account for the difference: (i) the survey design – VNLSS was a cross-sectional one while our study used follow-up surveys. The validity of data collected in FilaBavi during 1999-2000 by four different methods: quarterly household follow-up surveys, household census, communal population register system, and neighborhood survey was studied by Huy et al.[19] Compared to the other methods used, the quarterly household follow-up survey was the best method for death registration. This would be also true for illness registration (ii) the gap between the poor and the rich by early 1990 was not so wide compared to late 1990s. (iii) VNLSS 1993 used expenditure as an economic indicator while in our study a kind of composite economic indicator was employed. (iv) The quality of data guaranteed by means of adequate supervision, carried out regularly by field supervisors and research students. (v) Furthermore, the fact that people with better health status leave for cities or elsewhere to find jobs also result in higher rates of reported illness among remaining less healthier people in the study area.

Health Care Utilization and Health Care Expenditures

The study revealed several similar futures that were indicated in earlier studies carried out in the same setting using cross-sectional design.[20] Self-treatment was a common practice followed by visiting private practitioners. No-treatment was reported more often in poorer quintiles in the study area. This study showed that people in richer quintiles used private practitioners and hospitals more often than the poor did. Likewise, the fact that the rich used higher level of care like policlinics and hospitals more often than the poor was also shown by other studies in Vietnam.[21] [22] Recent evidence from a number of developing countries has also shown that the demand for hospital-based or private provider-based health services is highly related to income. After controlling for other factors, patients demand more for higher-priced health services as their incomes increase.[23] Inequity in utilization of health services between men and women could be only clearly demonstrated in using hospitals and self-

18 General Statistic Office. Vietnam Living Standard Survey (VNLSS) 1993. Hanoi: Statistical Publishing House; 1993.

19 Tran Quang Huy, Nguyen Hoang Long, Dinh Phuong Hoa, Peter Byass and Bo Eriksson. 'Validity and Completeness of Death Reporting and Registration in a Rural District of Vietnam'. Scandinavian Journal of Public Health. (In prep).

20 ND Khe, NV Toan, LTT Xuan, B Eriksson, B Höjer, VK Diwan. Primary health concept revisited: where do people seek health care in a rural area of Vietnam? Health Policy 2002; 61:95-109.

21 General Statistic Office. Vietnam Living Standard Survey (VNLSS) 1993. Hanoi: Statistical Publishing House; 1993.

22 General Statistic Office. Vietnam Living Standard Survey (VNLSS) 1998. Hanoi: Statistical Publishing House; 1998.

23 N.V. Toan. 'Utilization of health services in a transitional society: studies in Vietnam 1991-1999'. PhD thesis. Stockholm:Karolinska Institutet; 2001.

medication. A possible explanation for minor gender differences in Vietnam is the fairly even level of education of women and men. Also, women have paid work to a relatively high degree.

Like in many countries, health user fees have now become commonly applied in Vietnam and played important part in creating revenue for the health care system. According MOH data in 1998, the share of out-off pocket payment is about 80% of total health expenditure.[24] Not surprisingly, our results showed that payments for health care services were substantial for households and households in the poorest quintile. The proportion of expenditure for health of households in poorest quintile was significantly higher than that of households in richest quintile. Our findings are similar to that of other studies done in Vietnam as well as elsewhere and the reasons for charging the poor more were also discussed in earlier studies.[25] [26] [27]

Considering the rapid social and economic development, what will happen to the people of Vietnam in the future in terms of health care equity? Our data point at the need of better understanding of not only the determinants of self-treatment, but also its health and economic consequences. Given the reasons for such practice, the current situation of declining quality in public facilities, the introduction of user fees at public facilities, the widespread availability of private services, particularly that of private pharmacies call for special measures. One feasible measure may be controlling the dispensing of some types of drugs, such as antibiotics without a prescription, in order to avoid the worst consequences of self-medication. The pattern of health service utilization particularly with regard to hospital services, which are the most heavily subsidized services and consumed by the higher income group of the population, urges for the need to make public subsidies available for a more efficient and equitable health care. The need of an effective social financing mechanism for the poor must be emphasized. Expansion of Vietnam Health Insurance with partly subsidized premiums for the low-income group is one option for maintaining access to health care in the presence of increasing fees.

24 Ministry of Health. Health statistics year book 1998. Hanoi: Medical Publishing House; 1998.
25 N.D. Khe, N.V. Toan, L.T.T. Xuan, B. Eriksson, B. Höjer and V.K. Diwan. Primary health concept revisited: where do people seek health care in a rural area of Vietnam? Health Policy 2002; 61:95-109.
26 Knowles, J., Behrman, J.R., Diokno, B.J. and McInnes, K. 'Key issues in the financing of Vietnam's social services'. Final report to the Government of Vietnam and the Asian Development Bank; 1996.
27 Ensor, T., San, P.B. 'Access and payment for health care: the poor of northern Vietnam'. Int J Health Plann Managem 1996; 11:69-83.

Annex 8.1a   Age standardized mortality rate

		Wealth Index Quintiles					Total
		I	II	III	IV	V	
Male	Rate	11.57	8.85	8.32	6.91	6.73	7.67
	95% CI	9.54-14.03	7.45-10.53	7.09-9.77	5.84-8.17	5.73-7.92	7.10-8.28
Female	Rate	3.73	3.97	4.56	4.20	3.58	3.81
	95% CI	2.79-4.99	3.11-5.08	3.68-5.64	3.39-5.20	2.88-4.45	3.43-4.24
Both	Rate	5.92	5.85	6.08	5.31	4.75	5.30
	95% CI	4.97-7.06	5.06-6.78	5.33-6.94	4.64-6.08	4.16-5.44	4.97-5.65

Annex 8.1b   Age standardized morbidity rate

		Wealth Index Quintiles					Total
		I	II	III	IV	V	
Male	Rate	5.62	5.03	4.76	4.48	4.00	4.62
	95% CI	5.52-5.73	4.95-5.10	4.69-4.82	4.42-4.54	3.94-4.05	4.59-4.65
Female	Rate	7.38	6.24	5.69	5.49	4.84	5.76
	95% CI	7.28-7.48	6.16-6.32	5.62-5.77	5.42-5.56	4.78-4.90	5.73-5.80
Both	Rate	6.50	5.63	5.23	4.98	4.42	5.24
	95% CI	6.43-6.57	5.58-5.69	5.18-5.28	4.94-5.03	4.38-4.46	5.22-5.26

Annex 8.2   Economic status of households was described by following variables:

a) Housing: 'roof', 'wall', 'floor' (categorical), 'electricity' and 'household floor area', 'number of persons in household' (quantitative).
b) Sanitary condition (categorical): 'water source', 'latrine', 'bathroom'.
c) House accessing (binary): 'bicycle', 'radio', 'refrigerator', 'TV', 'video', 'electric fan', 'sewing machine', 'wardrobe', 'telephone', 'motorcycle', 'car', 'agriculture machine', 'buffalo – cow' and 'other'.
d) Land area (quantitative): 'for rice production', 'for gardening', 'for industry production' and 'for forest cultivation'.
e) Annual income (quantitative): 'agriculture', 'breeding', 'forestry', 'handy craft making', 'fishing', 'small service', 'supported by other people', 'small trading', 'salary' and 'other'.
f) Expenditure:
    • Daily food expenditure (quantitative): 'rice', 'meat', 'fish', 'eggs', 'vegetable' and 'other'.
    • Monthly expenditure (quantitative): 'buying valuable item', 'health care', 'education', 'fertilization', 'relationship' and 'other'.
g) Debt:
    • 'Amount of debt' (quantitative).
    • Reasons of debt (binary): 'daily expenditure', 'health care', 'education', 'reproduction', 'relationship', 'shopping' and 'other'.

Chapter 9

Does Health Intervention Improve Health Equity? Evidence from Matlab, Bangladesh

Abdur Razzaque and Peter Kim Streatfield

Summary

Two population-based birth cohorts of 10 years apart were followed for 5 years to examine the socioeconomic and gender inequality in mortality in the ICDDR, B-service and government-service areas of Matlab. The under-5 mortality declined more over time (% decline) in the ICDDR, B-service than in the Government service area and the same is true for infant but for 1-4 years, the declined was almost similar. Over the period, poor-least poor ratio of under-5 mortality had widened in both the areas while girls had higher mortality than boys in the earlier cohort, particularly in the Government service area but such difference disappeared in the recent cohort. If under-5 mortality is disagregated, poor-least poor mortality ratio had widened for infant in both the areas while the same is true in the Government service area for 1-4 years but the ratio declined considerably in the ICDDR, B-service area. More boys died during infancy than girls in both the areas and in each cohort but for 1-4 years, more girls died than boys in the earlier while such difference disappeared in the recent cohort.

Under-5 children of poor households died more due to diarrhoea than those of least poor households in both the areas and the same is true for pneumonia in the Government service area but poor and least poor died almost equally in the ICDDR, B-service area.

Background

Like many other developing countries in the world, mortality declined appreciably in Bangladesh since the middle of last century, however, under-5 mortality declined remarkably, particularly during the last two decades. In fact, such improvement in mortality had happened when the economy of the country was poor (per capita GDP US$ 280), with about 80% people living in rural areas, and about 60% were illiterate. Malnutrition is widespread and with about 40% of babies born annually classified as low birth weight (less than 2.5 kg), however, success in family planning resulted in a decline in total fertility rate, from 6.5 in the early 1970s to 3.4 in the mid-1990s.

To improve health of the general people, the government of Bangladesh has undertaken different programs since Alma Ata Conference in 1978, however, emphasis was often given to improve health of the mother and child. In half of the Matlab Demographic Surveillance System (DSS) area, the International Centre for Diarrhoeal Disease Research, Bangladesh (ICDDR, B) has been maintaining a Maternal Child Health and Family Planning (MCH-FP) intervention (ICDDR, B-service area) since late 1977, while the other half of the area is provided with usual government service (Government service area). However, these interventions were made available to all people, not targeting to the poor.

In the past (early 1980s), few studies documented lower mortality levels among children of higher socioeconomic than lower socioeconomic status, and higher mortality among girls than boys. Using Matlab DSS data, D'Souza and Bhuiya (1982) documented a negative relationship in mortality between children aged 1-4 and various socioeconomic factors. An examination of cumulative life-table probabilities of mortality by sex for the cohort born during 1973-75 in Matlab area revealed that in ages 1-4, mortality for girls exceeded that for boys by 59% (Koenig and D'Souza, 1986).

In recent years (since mid-1990s), a few studies have examined the effect of health interventions on socioeconomic inequality in mortality but results are not conclusive. Using Matlab DSS data, Muhuri (1995) reported that the health intervention program had a greater effect on the risks of child death (1-4 years) of uneducated mothers than mothers with at least some education. But following re-analysis of Matlab data for the same period, Razzaque (2002) did not find such educational effect on child mortality. In another study in Matlab, Bhuiya et al. (2001) reported that the poor-least poor difference in 1-4 years mortality had declined over time (1982-96).

The objective of the study is to examine socio-economic and gender inequalities in under-5 mortality in the ICDDR, B-service and Government service areas of Matlab, Bangladesh. These two areas are of similar socioeconomic condition but differ in quality of health and family planning services. Specifically, the study examines whether health intervention programs reduce socio-economic and gender inequalities in under-5 mortality and whether the inequalities changed overtime.

Methods

Setting

Data for this study come from Matlab *Upazila* (sub-district) where the International Centre for Diarrhoeal Disease Research, Bangladesh (ICDDR, B) has been maintaining a field station since 1963. Matlab is a rural area located about 55 km south-east of Dhaka. The area is low-lying deltaic plain intersected by the tidal river Gumti and its numerous canals. The major modes of transport within the area are walking, country boat and in some cases small steamer or launch. Farming is the dominant occupation, except in a few villages where fishing is the means of livelihood. Most of the farmers are in marginal situations with less than two acres of land, and 40% of them are landless. For many families sharecropping and work on others' land on a daily wage basis have become the main sources of livelihood. Some people also work in mills and factories in different towns and cities but their families live in the study area. Rice constitutes the staple food and is harvested three times annually. Rates of illiteracy are high and increase with age.

In Matlab study area the ICDDR, B has been maintaining a Demographic Surveillance System (DSS) over 200,000 population since 1966. The HDSS collects information on births, deaths, migration, marriages, divorce and household splits. The HDSS events are collected by the Community Health Research Worker (CHRW) through monthly household visits and Field Research Supervisor (FRS) are supervisor (in the past, CHRW recorded events through fortnightly household visits and FRS accompanied by the CHRW visited the household every six weeks to complete the registration form). The DSS also maintains cross-sectional socioeconomic data and such data is available for 1974, 1982 and 1996.

The Matlab HDSS (DSS named as HDSS after integrating with health data) area consists of ICDDR, B-service (population 107,369) and Government service (population 112,383) areas. Most of the ICDDR, B-service area was exposed to a contraceptive distribution program during 1975-77 and has been exposed to the Maternal Child health and Family Planning services since October 1977 (for details, see Bhatia et al., 1980). In the ICDDR, B-service area, MCH-FP services have been provided by the CHRWs through fortnightly home visits until December 1999 and then from fixed site clinics. In fact, the intervention began with the family planning services and all other health services have gradually been added over a period of time. These health services are: tetanus toxoid vaccination (pregnant women) – since March 1978, diarrhoeal disease management by *bari* mother – since October 1978, tetanus toxid vaccination to all mother (Blocks A+C) – since December 1981, measles vaccination to all children (Blocks A+C) – since March 1982 and to all children (Blocks B+D) – since December 1985, DPT and polio vaccination to all children (all 4 Blocks) – since March 1986, vitamin A distribution (all 4 Blocks) – since January 1987, maternity care (Blocks C+D) – since March 1987, ALRI detection and management with penicillin (Blocks B+D) – since April 1988, ALRI management with oral cotrimoxazole – since February 1990, maternity care (Blocks A+B) – since April 1990, antenatal check-ups introduced – since February 1991, BCC intervention to reduce ALRI mortality – since February 1991, maternal vitamin A supplementation after delivery – since March 1992, maternal supplementation with beta-carotene – since January 1993, reproductive tract infections detection and management – since June 1994, health centre assisted deliveries – since 1996 (Block C), since 1998 (Block D), since 2000 (Block B) and since 2001 (Block A), reproductive health program initiated since January 1997 (Van Ginneken et al., 1998; Koenig and Strong, 1995).

During the household visit in the ICDDR, B-service area, the CHRW asks mothers about their menstrual status, contraceptive use, contraceptive-related side-effects, pregnancy, breastfeeding and morbidity. The CHRWs are used to provide contraceptives and basic medicines to mother and child and referred patients with complications to sub-centre clinics (until December 1999). Currently services are being provided from the fixed sites (residence of CHRWs). In the ICDDR, B-service area there are four sub-centres that also provide MCH-FP services. In addition, ICDDR, B has a free 60-bed diarrhoea treatment centre in Matlab town and the facility is used not only by Matlab people but surrounding districts.

In the non-ICDDR, B-service area (Government service area), the government supported health service was available mainly in the urban areas until mid-1970s. The services were more curative than preventive in nature. The government of Bangladesh accepted the primary health care concept as national health objective in 1978. Since then, the health care system was reoriented to provide essential health care to the general mass. In fact, the government significantly increased the funding of health sectors from early 1980s. The facilities include Maternal and Child Welfare Centre in urban and sub-urban areas, Upazila Health Complex at Upazila level and the Family Welfare Centre at union level (DGHS, 1990). The government has also made primary health service facilities available at Rural Dispensaries and Satellite Clinics. In Matlab town, the government runs a 30-bed free general hospital along with few union-level health facilities. In addition, the government has been promoting oral rehydration therapy for diarrhoea management, and the immunization program against six major childhood diseases (Huq, 1991). In fact, immunization against childhood diseases is mainly delivered through satellite clinics (since mid-1980s) while family planning services are delivered at the door-step until recently.

As health services are added gradually both to the ICDDR, B-service and Government service areas, two birth cohorts (1983-85 and 1993-95) under study benefited differently from these interventions. For example, measles vaccination to all children started in 1982 in two blocks of the ICDDR, B-service area while such vaccination started in 1985 in other two blocks. So, the earlier birth cohort (1983-85) benefited less from these interventions than the subsequent birth cohort (1993-95). On the other hand, health services were also been added gradually in the Government service area, however, such service is less intensive than the program of the ICDDR, B-service area.

Data

The study used HDSS data from both the ICDDR, B-service and Government service areas. Two population-based birth cohorts of 10 years apart (1983-85 and 1993-95) were selected for analysis. The HDSS system registered 20,665 births for cohort 1983-85 and 16,925 births for cohort 1993-95 and these births were followed for five years for death and out-migration to ascertain survival and migration status respectively. The birth cohort of 1983-85 was matched with the socio-economic data of 1982 and birth cohort of 1993-95 was matched with socio-economic data of 1996 to copy the socio-economic data in the birth file.

The socio-economic status is defined here in term of assets, rather than income or consumption. The asset information was collected through the household questionnaire administered during the censuses. These questions include ownership of a number of consumer items (radio, watch, etc), dwelling characteristics (wall and roof material), type of drinking water and toilet facilities, however, more consumer items were collected in 1996 than in 1982 census. Data on causes of death were collected and coded by the FRS until 1986 and since then it is being collected by the FRS but coded by the medical assistant using modified version of ICD-9 codes (until 2002).

Data analysis

Economic status of the household is measured by constructing a wealth index using asset ownership as validated by Filmer and Pritchet (1998). Each household asset for which all information collected was assigned a weight or factor score generated through principal components analysis. The resulting asset scores were standardized in relation to a standard normal distribution with a mean of zero and a standard deviation of one. Each household was assigned a standardized score for each asset, where the score differed depending on whether or not the household owned that asset.

The equity gap can be measured in absolute (difference) or relative (ratio) scales. Use of absolute scales may often lead to apparent reductions in inequity gaps, because baseline rates that are already low (for least poor) are unlikely to decrease in absolute terms as fast as those are already high (for poor). Ratio scales, on the other hand, take into account the different baseline levels, and are thus more appropriate for examining time trends. In this study, ratio scale was used to examine the inequalities. Life table mortality rates were calculated for infant, 1-4 years and under 5.

Results

Table 9.1 shows mortality rates by cohort, study area and age. Over the study period, under-5 mortality declined in both the areas, and the decline was more marked in the ICDDR, B-service than the Government service area (46% vs 36%). The differential decline of under-5 mortality between the ICDDR, B-service and Government service area is due to difference in infant mortality (42% vs 27%) while for 1-4 years mortality, the decline was almost same in these two areas (57% vs 55%).

The level of mortality was lower in the ICDDR, B-service than in the Government service area for each age categories in both the cohorts (Table 9.1). In the earlier cohort, mortality rates were 19%, 13% and 30% lower in the ICDDR, B-service than in the Government service area for under 5, infant and 1-4 years respectively while comparable figures for the recent cohort were 32%, 31% and 34%. In fact, the mortality difference between the two areas has widened over time for under 5 (19% to 32%) and infant (13% to 31%) but the difference remains almost same for 1-4 years (30% to 34%).

Table 9.2 shows mortality rates by cohort, study area and wealth index. Under-5 mortality declined more over time among least poor than the poor in both the ICDDR,

Table 9.1   Mortality rates (per 1000) by cohort, study area and age

Cohort	ICDDR, B-service			Government service			ICDDR, B:Govt.		
	Infant	1-4 yr	Under 5	Infant	1-4 yr	Under 5	Infant	1-4 yr	Under 5
1983-85	100.9	40.9	141.8	116.5	58.7	175.2	0.87	0.70***	0.81**
1993-95	58.7	17.4	76.1	85.5	26.5	112.0	0.69**	0.66	0.68**
% decline	42*	57*	46*	27**	55*	36*	–	–	–

* p<0.001,   ** p<0.05,   *** p<0.10

Table 9.2   Mortality rates (per 1000) by cohort, study area and wealth index

Cohort	ICDDR, B-service					Government service				
	Poorest	Second	Middle	Fourth	Least poorest	Poorest	Second	Middle	Fourth	Least poorest
Infant										
1983-85	103.8	104.9	98.5	109.9	93.0	131.4	116.3	121.8	108.5	106.2
1993-95	67.1	66.8	67.5	52.0	43.7	97.7	88.0	95.6	66.8	63.4
% decline	35**	36**	31**	53*	53*	26**	24**	22***	38*	40*
1-4 yr										
1983-85	53.5	46.8	40.2	46.2	25.1	80.4	76.3	55.4	43.4	46.4
1993-95	20.6	21.3	19.8	15.6	10.9	39.1	34.2	20.5	18.4	17.3
% decline	61*	54*	51*	66*	57**	51*	55*	63*	58*	63*
Under 5										
1983-85	157.3	151.7	138.7	156.1	118.1	211.8	192.6	177.2	151.9	152.6
1993-95	87.7	88.1	87.3	67.6	54.6	136.8	122.2	116.1	85.2	80.7
% decline	44*	42*	37*	57*	54*	35*	37*	34*	44*	47*

* p<0.001,   ** p<0.05,   *** p<0.10

B-service (54% vs 44%) and Government service area (47% vs 35%). If under-5 mortality is disaggregated for infant and 1-4 years, such pattern (least poor declined more than poor) exist for infant in both the areas (53% vs 35% in ICDDR, B-service and 40% vs 26% in Government service) but for 1-4 years, mortality of least poor and poor declined almost equally in ICDDR, B-service area (57% vs 61%) while not in Government service area (63% vs 51%).

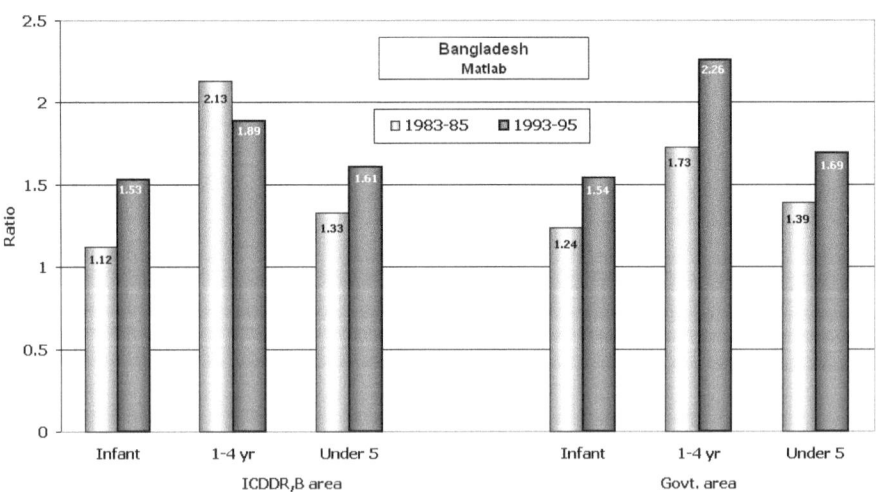

Figure 9.1   Poor-least poor ratio of mortality by study area, age and cohort

In the ICDDR, B-service area, poor-least poor difference of under-5 mortality exist in both the earlier and recent cohorts and poor-least poor mortality ratio has widened overtime (1.3 to 1.6) (Figure 9.1). A similar pattern also existed in the Government service area, poor-least poor difference in the earlier and recent cohorts and poor-least poor mortality ratio has widened overtime (1.4 to 1.7). In the ICDDR, B-service area, the increase in poor-least poor ratio of under-5 mortality is mainly due to increase in difference of infant mortality. In fact, poor-least poor ratio of infant mortality increased from 1.1 to 1.5 while poor-least poor ratio of 1-4 years mortality declined slightly (2.1 to 1.9). In the Government service area, the increase in poor-least poor ratio of under-5 mortality is due to increase in poor-least poor ratio of both infant and 1-4 years mortality. In fact, poor-least poor ratio of infant mortality increased from 1.2 to 1.5 while poor-least poor ratio increased from 1.7 to 2.3 for 1-4 years mortality.

Table 9.3 shows mortality rates by cohort, study area and sex. Under-5 mortality declined more over time among girls than boys in both the ICDDR, B-service (49% vs 44%) and Government service (42% vs 29%) areas. If under-5 mortality is disagregated for infant and 1-4 years, survival improved more among girls than boys for 1-4 years (61% vs 38% in ICDDR, B-service and 63% vs 41% in Government service area) but during infancy, survival improved almost equally among girls and boys (43% vs 45% in ICDDR, B-service and 29% vs 25% in Government service area).

The boy-girl mortality difference exists in the earlier but such difference had disappeared in the recent cohort for ICDDR, B-service (ratio, 0.9 to 1.0) and Government service (ratio, 0.85 to 1.05) area (Figure 9.2). The elimination of a boy-girl difference of under-5 mortality is due to reduction in boy-girl difference of 1-4 years mortality (ratio, 0.53 to 0.84 in ICDDR, B-service and 0.56 to 0.89 in

Table 9.3   Mortality rates (per 1000) by cohort, study area and sex

Cohort	ICDDR, B-service		Government service	
	Boy	**Girl**	**Boy**	**Girl**
Infant				
1983-85	108.8	100.1	119.0	113.9
1993-95	60.1	57.1	89.8	81.1
% decline	45*	43*	25**	29**
1-4 yr				
1983-85	26.0	48.8	42.2	76.0
1993-95	16.0	19.1	25.0	28.1
% decline	38	61*	41**	63*
Under 5				
1983-85	134.8	148.9	161.2	189.9
1993-95	76.1	76.2	114.8	109.2
% decline	44*	49*	29**	42*

* p<0.001,   ** p<0.05,   *** p<0.10

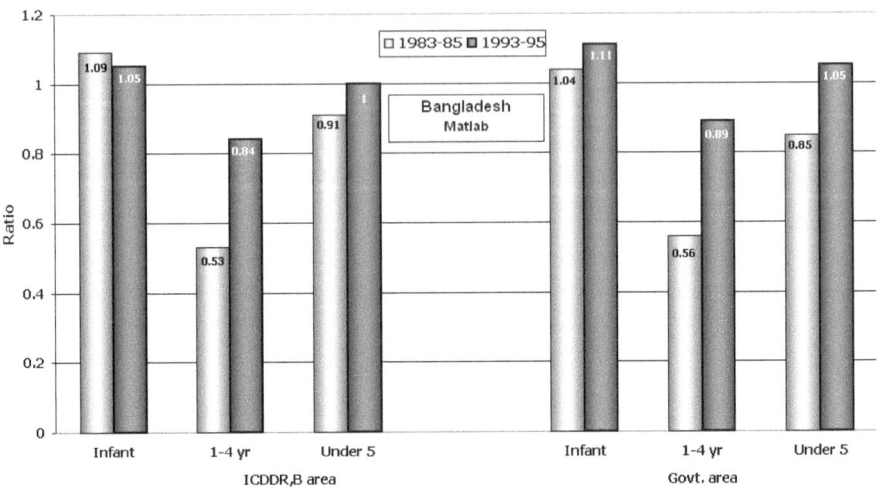

Figure 9.2   Boy-girl ratio of mortality by study area, age and cohort

Government service area) while during infancy more boys died than girls (ratio, 1.09 to 1.05 in ICDDR, B-service and 1.04 to 1.11 in Government service area).

Table 9.4 shows under-5 mortality rates by cause, wealth index and study area, cohort (83-85) and cohort (93-95). The poor had higher mortality than least poor for diarrhoea and other causes in both the areas and in both the cohorts but for pneumonia death, poor-least poor difference did not exist in the ICDDR, B-service area while it exist in the Government service area (Figure 9.3).

Table 9.4   Mortality rates* (per 1000) by cause of death, wealth index and study area

Cause of death	ICDDR, B-service					Govt.-service				
	Poorest	Second	Middle	Fourth	Least poorest	Poorest	Second	Middle	Fourth	Least poorest
Cohort 83-85										
Diarrhoea	33.3	35.7	29.4	31.9	24.0	49.5	48.2	36.4	24.9	28.4
Pneumonia	15.4	15.9	14.2	15.2	14.5	16.5	16.8	19.5	15.9	13.5
Others	105.6	98.1	94.0	106.1	78.6	138.9	122.9	118.2	108.8	107.2
Cohort 93-95										
Diarrhoea	14.1	11.8	15.3	9.1	6.2	22.4	19.7	14.9	9.9	13.2
Pneumonia	8.9	14.1	19.3	6.5	11.0	26.1	23.2	22.9	21.0	9.7
Others	64.0	61.4	51.9	51.1	36.5	86.8	77.9	77.4	52.5	56.2

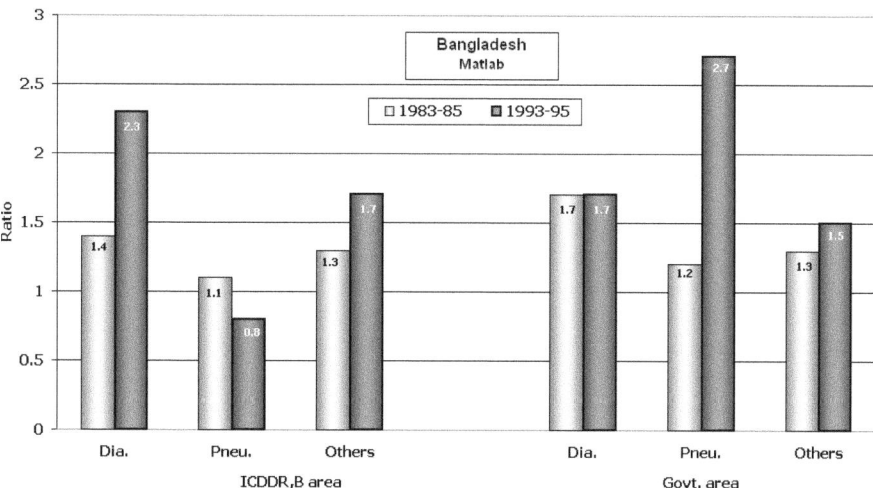

Figure 9.3   Poor-least poor ratio of mortality by study area, death cause and cohort

Discussion

Before interpreting results, the following points should be kept in mind. To measure the impact of health intervention on time trend data is complex because some interventions might reach the poor after a certain period but at the same time a new intervention would benefit more least poor than the poor. Although it is usually believed that the poor would benefit more from health interventions than the least poor, but such interventions are not usually targeted for the poor.

Over the 10 years period, under-5 mortality declined substantially in both the areas and declined more in the ICDDR, B-service than in the Government service area. Such pattern has widened mortality difference in the two areas and is due mainly to improvement of infant survival. In fact, infant survival improved more in the ICDDR, B-service than in the Government service area due to the nature of intervention that affected neonates (antenatal care, supply of delivery kits etc); these interventions have been in operation in the ICDDR, B-service area since early 1990s. On the other hand, mortality of 1-4 years was at a relatively low level for the earlier cohort in the ICDDR, B-service area and it was because of the childhood interventions (immunization). These services were available in the ICDDR, B-service area (two Blocks) since early 1980s while such services were introduced in the Government service area in late 1980s (EPI program).

The poor-least poor difference of under-5 mortality had widened over time in both the areas and it is due to the difference in infant mortality, however, such pattern was absent for 1-4 years in the ICDDR, B-service area where the poor-least poor mortality difference had reduced considerably. While health services are made available, the least poor initially use more than the poor, however, such services can reach to other economic class if it is intensive as found in case of childhood immunization (1-4 years) in the ICDDR, B-service area and probably has reduced mortality difference between poor and least poor. To improve infant survival, interventions were introduced at later date (early 1990s) and it was used more by the least poor than the poor that might increased poor-least poor differences in infant mortality. A similar pattern has also been observed in Brazil, where Victora et al. (2000) reported that a new intervention was initially adopted by the wealthy and increased the mortality gap, but subsequently the gap was reduced when that intervention was adopted by the poor.

Although more under-5 girls died than boys in the earlier cohort in both the areas, such difference had disappeared in the recent cohort. However, girls benefited more over time than the boys among aged 1-4 years, while girls and boys are equally benefitted among infants. In recent years, a further decline in desired family size has also been observed (4.5 in mid-1970s to 2.5 in mid-1990s) and it had happened after actual fertility started declining. Over the period, motivation for small family size has increased and couple in these days prefers to have quality-children rather than quantity (Razzaque, 1996). As desired family size is reaching close to replacement level, son preference is declining and parents probably are treating both boys and girls almost equally as found in secondary school attendance (Razzaque and Streatfield, 2001). However, in the past it was reported that girls were discriminated against in terms of intrafamily food distribution, medical care, education and so on (Chen et al., 1981; D'Souza and Chen, 1980).

The poor died more from diarrhoea than least poor in both the areas, but for pneumonia such a pattern exists only in the Government service area. Such poor-least poor differences in diarrhoea mortality is unexpected because *bari* mothers (one in 15 households) have been providing ORS to diarrhoea patients since 1978 in the ICDDR, B-service area and the service is expected to reached both poor and least poor equally. However, use of oral saline during diarrhoea is low in both the ICDDR, B-service and Government service areas (45% vs 35%) and poor-least poor differences in use did not exist (Razzaque et al., 2002). For pneumonia, poor and least poor died equally in the ICDDR, B-service area and it is probably because of the service provided by ICDDR, B (oral cotrimoxazole since 1990). However, medicine use during pneumonia is high in both the areas (about 80%) and poor-least poor differences do not exist (Razzaque et al., 2002).

Although under-5 mortality declined over time in both the areas but poor-least poor differences had increased while boy-girl differences disappeared in the recent cohort. Although the government gives priority to improving health of the people but the programs usually not designed to improve socioeconomic and gender inequality in mortality. The findings of the study are encouraging because mortality is declining in both the areas, along with improvement of gender equity. Moreover, there is a possibility that the poor-least poor difference in mortality would reduce in the future once the least poor achieves low mortality level, however, it may take a long time, even in the intensive program area, as observed among children (1-4 years) in the ICDDR, B-service area. However, such intensive health interventions may be difficult to operate in a natural environment (government program) where the poor-least poor difference is still increasing.

References

Bhuiya, A., M. Chowdhury, F. Ahmed and A.M. Adams (2001). 'Bangladesh: An Intervention Study of Factors Underlying Increasing Equity in Child Survival'. In Evans et al. (ed), *Challenging inequities in health: from ethics to action*. New York: Oxford University Press, pp 227-39.

Bhatia, S., W.H. Mosley, A.G. Faruque and J. Chakraborty (1980). 'The Matlab family planning – health services project'. *Studies in Family Planning*, 11(2):202-211.

Chen L.C., E. Huq and S. D'Souza (1981). 'Sex bias in the family allocation of food and health care in rural Bangladesh'. *Population and Development Review*, 7(1):55-70.

D'Souza S. and A. Bhuiya (1982). 'Socioeconomic mortality differentials in a rural area of Bangladesh'. *Population and Development Review*, 8(4):753-769.

D'Souza S. and L.C. Chen (1980). 'Sex differentials in mortality in rural Bangladesh'. *Population and Development Review*, 6(4): 753-769.

DGHS (Directorate General of Health Services, 1990). *Bangladesh Health Services 1989*. Government of the People's Republic of Bangladesh, 1990.

Huq, Mujibul (1991). *Near Miracle in Bangladesh*. Dhaka University Press Limited.

Koenig, M.A. and S. D'Souza (1986). 'Sex differentials in childhood mortality in rural Bangladesh'. *Social Science and Medicine*, 22(1):15-22.

Koenig, M.A. and M. Strong (1995). 'Assessing the mortality impact of an integrated health programme: Lessons from Matlab, Bangladesh'.

Muhuri, P.K. (1995). 'Health programs, maternal education and differential child mortality in Matlab, Bangladesh'. *Population Development Review*, 21(4): 813-834.

Pritchett, L. and D. Filmer (1998). 'Estimating wealth effects without expenditure data or tears: with an application to education enrolment in states of India'. World Bank Policy Research Working Paper No. 1994, October 1998.

Razzaque, A (unpublished) (2002). Results from HDSS database, ICDDR, B.

Razzaque A. 'Reproductive preferences in Matlab, Bangladesh: levels, motivation and differential', *Asia-Pacific Population Journal*, 11(1): 25-44, 1996.

Razzaque, A. and P. Kim Streatfield (2001). 'Family size and children's education in Matlab', Bangladesh. (unpublished).

Razzaque, A. and P. Kim Streatfield (2002). 'Health care use and health equity in Matlab', Bangladesh. (unpublished).

Van Ginneken, J., R. Bairagi, A. de Francisco, A.M. Sarder and P. Vaughn (1998). 'Health and Demographic Surveillance System in Matlab: Past, Present and Future'. Special Publication No. 72, ICCDR, B Dhaka.

Victora, C.G., F.C. Barros and J.P. Vaughan (2000). 'The impact of health interventions on inequalities: infant and child health in Brazil'. In Leon and Walt (eds), *Poverty inequality and health: An international perspective*. Oxford University Press.

Chapter 10

Socio-economic and Regional Disparity in the Utilization of Reproductive Health Services in Bangladesh

Abdullahel Hadi and M. Showkat Gani

Summary

Although the health care system has significantly expanded in Bangladesh during the last two decades, the health status of the population has remained very poor because of the uneven distribution of services. Inequality in health exists in many forms and multiple dimensions such as age, sex, education, income, ethnicity, etc. Using data from a nationally representative sample, this study attempts to improve our understanding about the socioeconomic and regional disparity in the utilization of reproductive health services in Bangladesh.

Data for this study came from the demographic and health surveillance system of BRAC which provided the updated information of the ownership of household asset and the use of reproductive health services. Socioeconomic disparity was measured by constructing a wealth index using compound assets and possessions of a set of household wealth. The surveillance areas were categorized into four regions as urban slum, rural under-served, other rural and the hill tracts. The utilization of reproductive health services was measured by the use of ante and postnatal care, maternal immunization coverage, and the use of safe delivery. A total of 1,182 randomly selected women, who gave birth in 2001, were interviewed.

Findings revealed significant socioeconomic and regional differentials in the use of reproductive health services. The use of services was much lower among the extreme poor than the non-poor and among the ethnic minorities in the hill and rural under-served than the other regions. The region specific inequalities, which were greater than the socioeconomic inequalities, may be reduced by expanding outreach health programs to bring services closer to the disadvantaged. The study concludes that much of these inequalities are social constructs that can be reduced by prioritizing the needs of the disadvantaged and adopting appropriate policy change options.

Background

Equal opportunities for health are desirable goal in all societies. It is expected that everyone should have a fair chance to attain their full health potential and that none should be excluded from achieving this. Although the health status in most countries of the world has significantly improved over the past few decades, substantial inequalities in health outcomes among nations, socioeconomic groups and individuals have remained (Leon and Walt, 2001). Improving the health of the poor and reducing health inequalities have become the central goals of many development programs (Wagstaff, 2002). Four dimensions in health such as equal access to available care for equal need, equal utilization for equal need, equal quality of care for equal need, and equity in outcome are emphasized to promote health equity (Krasnik, 1996).

Studies revealed that poverty and ill-health are intertwined (Wagstaff, 2002) and that poverty and marginalization are the underlying causes of inequities in health (Evans et al., 2001). The health status of the poor requires to be understood by their social conditions including access to the basic needs and amenities like food, drinking water, housing, education, employment, transport and communication (Prasad, 2000). The socioeconomic well-being of the poorer section of the community has deteriorated in many developing countries in relative terms because the rural poor were generally denied access to resources needed to them by which they could improve their own incomes and living conditions (World Bank, 1993). The general process of economic growth which involves the increase in national income per capita may not necessarily promote health unless choices to be made on the priorities to be chosen (Sen, 1999).

The poor and women are expected to suffer a greater burden of ill health than do the rest of the population particularly during pregnancy and childbirth. The need of the expansion of reproductive health services in developing countries has now been recognized than ever. More than 500,000 maternal deaths that occur every year of which a quarter to a third of all deaths is the result of complications of pregnancy (WHO, 2000). The regional variation in reproductive health outcome is also very wide. More than 99% of maternal deaths occur in developing countries. A woman living in Africa has 200 times greater risk of dying from complications related to pregnancy than a woman living in an industrialized country (WHO, 2000).

Although the poor faces the worse reproductive health outcome than others, poverty is not an insurmountable barrier to health if appropriate investment in health is made. There are many discriminatory policies in place in most developing countries. One major problem in reaching the poor women has been that most cost-effective interventions are not targeted to the poor but targeted at the very diseases from which the poor suffer disproportionately (Feachem, 2000). Moreover, the rich receive more of subsidies than the poor and, as a result, the cost of health care deters poor from seeking care to a greater extent than the rich (Wagstaff, 2002). The distribution of public health services is unequal in many developing countries (Makinen et al., 2000). The use of health services with an illness or injury varies significantly by consumption quintiles and urban-rural differences (Baker and van der Gaag, 1993).

Bangladesh is a poor country with nearly half (48%) of the population lived on the wrong side of the poverty line. The poor are vulnerable to illnesses by a combination of low levels of education and poor access to health services. Thus, regardless of the increase of access to primary health care in recent years, the poor continue to suffer from lack of access to services. The community-based non-formal education has been promoted in some countries to raise health information and reduce morbidity with marked improvement in health knowledge in many countries (Power, 1996). Although the health care network has expanded in the rural areas of Bangladesh and the country has experienced significant health development over the past two decades, the overall livelihood situation of the poor women has changed very little.

The problems of health care system are deeply rooted in the society and their transformation requires major structural changes. Several attempts have been made to understand the equity issues in Bangladesh although, the questions to many issues have remained unanswered (Bhuiya et al., 2001; Chowdhury and Bhuiya, 1999). Using longitudinal cohort data, they have shown the existence of gender and socioeconomic inequality in child survival, intra-household food distribution, child nutrition, family planning, literacy and family violence, and demonstrated how intervention programs have promoted equity in health. Both studies were conducted in a small geographic area with a very homogeneous population in Matlab and, thus, the other dimensions of inequalities such as urban-rural and regional differentials in health have not been addressed. The variation in reproductive health status has consistently been reported very wide and has never been uniform across the country (Mitra et al., 1997). This study attempts to improve our understanding about the socioeconomic and regional disparity in the utilization of reproductive health services in Bangladesh. Four domains of reproductive health services viz. antenatal and postnatal care, maternal immunization and safe delivery were considered in this study.

Methods

Data Source

Data for this study came from *Watch Project*,[1] the demographic and health surveillance system of BRAC covering more than 90,000 people living in 85 villages in ten rural areas, four urban clusters in a large metropolitan city and five ethnic clusters in the hill regions of Bangladesh where several NGOs had poverty reduction programs since 1990s. The surveillance areas of Watch Project were selected to become representative of the country. In each surveillance area, BRAC operates a field research station to cover approximately 1,000 households in the neighbouring 6 to 8 villages. Two female and one male field investigators routinely visit all households each month and record relevant health and demographic information on the registers. The surveillance system provided updated sampling frame of all married women (aged <50 years) who had given birth in 2001. A systematic random sampling technique was followed to select sample women from the database. Data were collected by a team of women investigators who had professional training and experience in the survey research techniques. A structured questionnaire was used to collect detailed information of the socio-demographic characteristics of sample

---

1   *Watch Project* is the demographic and health surveillance system of BRAC covering more than 90,000 people living in 85 villages in ten rural areas, four urban clusters in a large metropolitan city and five ethnic clusters in the hill regions of Bangladesh. The surveillance system is considered representative of Bangladesh.

women, household wealth and use of reproductive health services. A total of 1,182 women were interviewed. Data were collected in April 2002.

Measuring Household Wealth Index

Watch Project had updated information about the household ownership of wealth such as table, cot, quilt, watch. radio, television and cycle. In addition, information about housing characteristics, use of electricity, source of drinking water, etc. were also available in the database. The approach in developing the household wealth index for this study has been developed by Filmer and Pritchett (2000) who have shown that the wealth index performs as well as a more traditional measure such as household-size-adjusted consumption expenditures. Following this approach, a set of household level variables was identified to include in the construction of wealth index.[2] These were table, cot, quilt, watch, radio, television, cycle and electricity. Each of the variables was recoded into categorical dichotomous (yes-no) variable. A total of 8 dichotomous variables was created and standardized. The principal component analysis was run with all constructed variables with certain criteria. The component score coefficient matrix was multiplied by the standardized variables to produce factor scores which were termed as household wealth score. The wealth scores were classified into quintile for this research.

Definitions of the Variables

The study focuses on the *utilization of reproductive health care*[3] as the outcome variable. Four aspects of reproductive health viz. the use of ante and postnatal care,

---

2   Steps to produce the wealth index: The socioeconomic information for each household of Watch Project areas were updated. A set of household level variables was identified to include in the construction of wealth index. These were table, cot, quilt, watch, radio, television, cycle and electricity. Each of the variables was recoded into categorical dichotomous (yes-no) variable. Thus, we created eight dichotomous variables. All variables were then standardized. The principal component analysis (factor analysis) was run with all constructed variables with the criteria as follows: only one factor to be produced, no rotation, principal components extraction, factor score to be calculated with regression method and print only component score coefficient matrix. The component score coefficient matrix was multiplied by the standardized (sampling weight) variables to produce factor scores which were termed as household wealth scores. The *household wealth* scores were classified into quintiles for this research.

3   Utilization of reproductive health services was measured by four indicators such as a) the use of antenatal and b) post-natal care, c) immunization coverage during pregnancy and d) delivery in the hospital. *Antenatal care* included participation of pregnant women in health education sessions, routine health check-up by a medically trained professional, monitoring weight gain, identification of danger signs, intake of iron tablets, vitamin and nutrition supplementation if needed and the use of emergency services for the complicated cases. If a woman had routine health check-up in the third trimester and took (iron and vitamin) supplementation, she was considered to have received antenatal care during pregnancy. *Post-natal care* covered follow-up visits by a medically trained professional, education on breastfeeding, food supplementation, counseling and post-partum period,

maternal immunization coverage, and safe delivery were examined in this study. All married women who had given birth in 2001 were asked to report about the relevant health services they received. If a woman had routine health check-up in the third trimester and took (iron and vitamin) supplementation, she was considered to have antenatal care during pregnancy. On the other hand, a woman was considered received post-natal care if she was monitored at least once by a health professional within 4 weeks after delivery. Similarly, a woman was considered immunized if she had both doses of tetanus toxoid vaccines. If she delivered in the health clinic or hospital under the supervision of a physician or midwife, the delivery was considered safe.

The socioeconomic and regional disparity in the utilization of reproductive health services was assessed in this study where the main independent variables were wealth index and region of residence. The other variables were age and education of women and ownership of land. Based on the preliminary analysis, *age* of women was dichotomized as <30 and 30+ years, *education* of women was coded as some or no education and *land ownership* was considered as continuous variable as decimals. *Region of residence*[4] was created from the widely dispersed and varied DSS sites. A total of four distinct regions such as hill, urban, traditionally under-served rural and other rural were identified based on certain criteria. The residents of the *hill region* in the south-east were primarily ethnic minorities who were very different than the mainstream population in terms of language, culture, religion, food habit and economic activities. The *urban* clusters consisted of urban poor and lower middle class in the city of Dhaka. *Traditionally under-served areas* included remote and inaccessible, and traditionally conservative villages where the public services were always found poor. The *other rural* sites were largely representative rural communities in Bangladesh.

---

growth monitoring and immunization for the newborn. If a woman was monitored at least once by a health professional within 4 weeks after delivery, she was considered to have received postnatal care in this study. The women were expected to receive two doses of tetanus toxoid (TT) during pregnancy. A woman was considered *immunized* if she had both doses of tetanus toxoid vaccines. When the newborn was delivered in a health clinic or *hospital*, the delivery was considered safe.

4   Region of residence: Watch Project coverage was expanded in 2002 to include ethnic minorities in the southeastern hill district of Bandarban, and urban poor and lower middle classes in the city of Dhaka. The ethnic minorities in the hill region were very different than the mainstream population in terms of language, culture, religion, food habit and economic activities. The urban clusters consisted of the lower classes and, therefore, not representative of urban population in general. Among the 10 rural sites covered by the *Watch Project*, two sites were located in the remote haor (riverine low land) and, as a result, were very inaccessible to the public facilities. Another two sites were located in traditionally conservative districts where the health services for women were consistently low and always found neglected (Mitra et al., 1997). These four sites constituted the traditionally under-served regions in this study. The remaining six rural sites were considered average and largely represented the rural communities of Bangladesh. Watch Project sites were, therefore, categorized into four regions such as hill, urban, rural under-served and other rural areas.

Results

Profile of Sample Women

The differences in socio-demographic characteristics of sample women by the region of residence are shown in Table 10.1. More than a third (35.7%) of the women was relatively younger with the mean age of nearly 28 years. The proportion of younger women were much larger in the hill and traditionally underserved regions compared to the urban and other rural regions. Illiteracy among women was widespread in the study areas as only 44.8% sample women could read and write. Literacy rate was significantly lower among the ethnic minorities in the hill compared to the other regions. Only 36% of the households had some kind of agricultural land. The mean ownership of land was only 1.6 hectares. As expected, the urban households had no farm land while the land ownership was highest in the hill regions. In summary, the socio-economic characteristics of sample women differ significantly by the region of residence.

Socio-demographic Factors and the Health Services Use

The utilization of all four components of reproductive health services was shown in Figure 10.1. About 36.5% pregnant women had received antenatal care while only half (18.2%) of them had accessed postnatal care. The maternal immunization coverage appeared to be very high (76.3%) compared to only 63.7% at the national average in 2000 (Mitra et al., 2001). The use of health facility such as hospital or clinic for the delivery, on the other hand, was only 4.1% compared to nearly 7.9% at the national level (Mitra et al., 2001).

   Table 10.2 shows the role of socioeconomic factors on the use of reproductive health services. Unlike other studies (Swenson et al., 1993), age of women appeared to have negative association with reproductive health service use although the

Table 10.1   Profile of sample women by region of residence

Study Variable	Hill[a]	Urban[b]	Region Rural-1[c]	Rural-2[d]	All
Age ≤30 years	41.7	27.4	43.6	27.4	35.7
Mean age (years)	30	28	30	27	28
Literacy rate	21.7	48.8	45.7	48.7	44.8
Land ownership	53.9	–	33.5	40.4	35.9
Mean amount of land (hec)	4.9	–	1.2	1.4	1.6
N	115	84	505	478	1182

[a] The residents of the hill region in the south east are primarily ethnic minorities.
[b] Urban samples include only the lower middle class and slum dwellers.
[c] Rural-1 represents the traditionally underserved and remote rural areas.
[d] Rural-2 represents other rural areas.

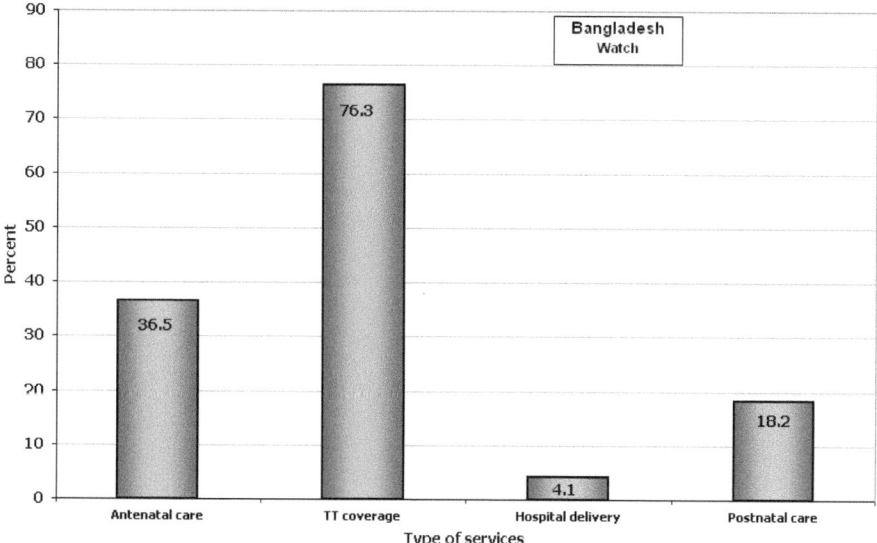

Figure 10.1   Utilization of reproductive health services by type

Table 10.2   Utilization of reproductive health services by socio-economic factors

Socio-demographic factors	Reproductive health services				N
	Received ANC	TT coverage	Delivery in hospital	Received PNC	
Age of women (years)					
≤ 30	40.8	79.1	4.5	19.1	422
30 +	28.9	71.3	3.3	16.6	760
P	.001	.003	.203	.287	
Education of women					
No education	26.5	71.0	2.3	13.2	652
Educated	48.9	82.8	6.2	24.3	530
P	.001	.001	.001	.001	
Land ownership (hectare)					
Landless	35.1	73.9	3.6	17.0	758
Land owner	39.2	80.7	5.0	20.3	424
P	.165	.009	.947	.163	

P-values are tests of heterogeneity.

relationships were not significant (at p<.05) for the hospital delivery and postnatal care. Negative association of maternal immunization and hospital delivery with age was also apparent in other studies as well (Magadi, Madise et al., 2000; Mitra et al., 2001). Education was more likely to positively influence the use of health care (Swenson et al., 1993; Kutzin, 2001; Mitra et al., 2001). Education of women, not only greatly strengthened their ability to understand the options available to them and modified their individual attitudes to seek care when in need but also increased their ability to make good use of health services (World Bank, 1993; Cook and Fathalla, 1996). Like education, land ownership was positively associated with the utilization of reproductive health services although the relationship was not statistically significant except for maternal immunization.

Economic and Regional Disparity in the Use of Services

The socioeconomic and regional disparity in the use of reproductive health was very wide (Table 10.3). The use of services in all four components significantly (p<.01) increased with the increase of household wealth. The differences in the utilization of reproductive health care by wealth index indicate that the extreme poor were significantly less likely to use antenatal health services than the moderate and other less poor. The picture was nearly similar for the case of postnatal care. The delivery in the hospital was nearly 8 times higher among the least than the extreme poor. This finding was consistent with other studies where the safe delivery rate was about 16.5

Table 10.3   Utilization of reproductive health services by wealth index and region of residence

Wealth index and region	Received ANC	TT coverage	Delivery in hospital	Received PNC	N
	Reproductive health services				
Wealth index					
Extreme poor	15.7	59.0	1.2	7.2	249
2	33.2	76.4	3.1	15.3	229
3	29.5	78.4	2.1	15.8	190
4	49.4	86.1	4.2	25.5	259
Least poor	52.2	81.6	9.0	25.9	255
P[a]	.001	.001	.001	.001	
Region					
Hill tracts	22.6	51.3	2.6	22.6	115
Urban	39.3	61.9	8.3	14.3	84
Rural underserved	27.3	68.7	2.6	17.0	505
Other rural	49.2	92.9	5.2	19.0	478
P[b]	.001	.001	.042	.386	

[a] P-values are tests of trend.
[b] P-values are tests of heterogeneity.

times higher among the rich than the poor (Gwatkin et al., 2000). About 82% of the least poor women were immunized during pregnancy compared with only 59% of the extreme poor women. The least poor – extreme poor inequality was less pronounced in TT coverage compared to the use of other reproductive health services.

The disparity in the utilization of reproductive health care by the region of residence was also very wide which might be related to regional differences in access to services (Swenson et al., 1993). Overall, the coverage of services was lowest among the ethnic communities in the hill than other three regions except for the use of postnatal care. The women in the rural underserved region had limited access because of the unavailability of the services and, thus, the utilization was poorer than other rural and urban regions. The women living in 'other rural' regions had better access than others in receiving antenatal care. As found elsewhere (Mitra et al., 2001), the use of tetanus toxoid during pregnancy and delivery in the hospital was much higher in the urban (8.3%) than other regions probably because of the better hospital facilities in the urban region.

Figure 10.2 shows that nearly 80% received at least one kind of reproductive health services. As expected, the coverage systematically reduced with the number of services. Only 2% women received all four types of reproductive health services in the study areas.

Table 10.4 shows the variation in the use of the number of health services received by wealth index and the region of residence. Although nearly 80% received at least one service (Figure 10.2), the proportion of women received at least one service was only 62.2% among the extreme poor. The use of services increased with the reduction of poverty level although the service utilization was reduced slightly among the least

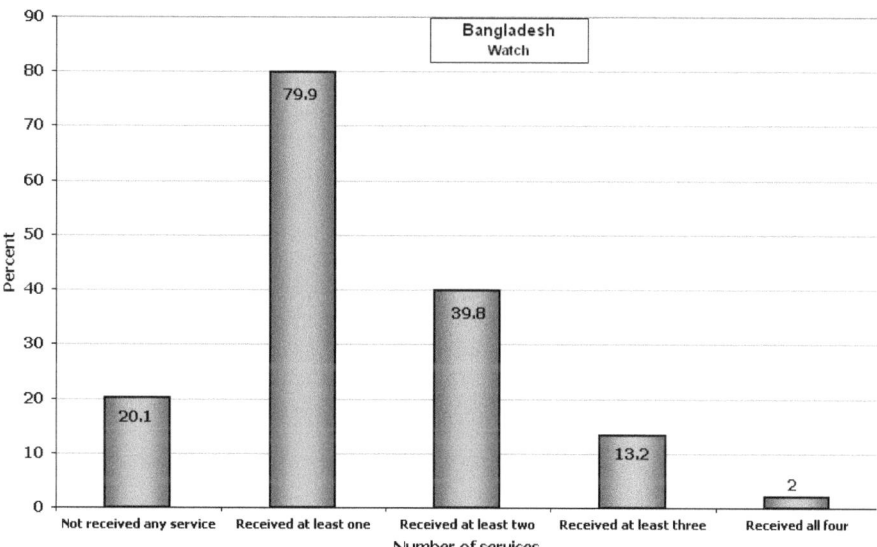

Figure 10.2   Use of reproductive health by the number of services

Table 10.4   Use of the number of reproductive health services by wealth index and region of residence

Wealth index and region	Not received	1	Received at least		
			2	3	4
Wealth index					
Extreme poor	37.8	62.2	16.8	3.9	0.4
2	19.7	80.4	35.3	10.4	0.9
3	19.5	80.5	34.7	9.4	1.1
4	9.7	90.3	54.4	17.3	2.7
Least poor	14.1	85.9	55.3	23.5	4.7
P		.001	.001	.001	
Region					
Hill tracts	40.0	60.0	24.3	13.9	1.7
Urban	35.7	64.3	40.5	16.7	3.6
Rural underserved	27.3	72.7	32.1	9.7	1.6
Other rural	4.8	95.2	51.7	16.3	2.3
P		.001	.001	.001	

[a] P-values are tests of trend.
[b] P-values are tests of heterogeneity.

poor who received at least one service. Relatively lower coverage of maternal immunization among the least poor (Table 10.3) appeared to be responsible for such unusual pattern of services. Among those who received at least two services, the coverage was positively associated with wealth index. When the coverage among extreme and least poor women were compared, the ratio was found to systematically increase with the number of services. Overall pattern of the use of services among the wealth quintiles indicates that wealthier groups have used health services more often when they need it compared to the poorer groups (Makinen et al., 2000).

As found earlier, women living in 'other rural' region had highest coverage among those who received at least one service while women living in the hill region were more deprived. This ratio varied significantly with the increase of the number of services. Among those who received three or more services, the coverage was highest in the urban region probably because of wider use of hospital delivery and post-natal care in the urban center. Overall, the reduction of coverage with the increase of the number of services was least among urban followed by the women of 'other rural' regions.

Discussion

This study provides two important findings: the use of reproductive health services was largely inadequate at the aggregate level and significant health sector inequality exists in Bangladesh. Although it is not certain whether the increased access to and the availability of services would lead to increased utilization of services among the poor and disadvantaged (Magadi, Madise et al., 2000), there were evidences which suggested that unavailability of health services not only reduced the coverage of services but also forced many women to seek alternative health care providers not acceptable by any standard (Whitehead et al., 2001). As found elsewhere, this study also shows that education of women not only raised their ability to understand the need of seeking care during pregnancy but also increased their ability to make good use of services (World Bank, 1993). The socioeconomic and regional disparity in the use of reproductive health services was very wide and the poor women living in the under-served region suffered a greater burden than others during pregnancy and post-natal period (Magadi, Madise and Rodrigues, 2000; Dowan and Brewster, 1998).

Although reproductive health services were expanded in the last two decades, it did not promote health equity because the services were available largely to urban centers. As a result, the use of reproductive health services has remained very low among the poor particularly among the ethnic minorities in the hills and under-served rural regions. Even in the urban and better-served rural areas, the poor-nonpoor disparity has continued to exist because the health services were not designed for the poor (Hadi et al., 2001).

Among the services, the coverage of maternal immunization was quite high compared to other components of reproductive health care although socioeconomic and regional differences in maternal immunization were also large. The high maternal immunization coverage indicates that reaching a larger group of pregnant women for other services was quite possible within the existing provision of MCH services. The use of post-natal services was about a half of that of the antenatal health services. While relatively poor showing during post-natal period was consistent with other studies (Mitra et al., 2001), this was not the case in the hill region. Although both the ante and post-natal health services were quite low in this group of ethnic minorities, a relatively better system of follow-up of pregnant women after delivery made it possible there to reduce the gap of coverage between the ante and post-natal services. The study clearly identifies the need of institutional delivery services particularly for the complicated pregnancies. The highest coverage of hospital delivery among the urban poor reflects the notion that the access to institutional delivery would raise the use of such services among the poor in other areas as well.

Disparity in enjoying public goods such as health care has never been new in Bangladesh. What is new is the recognition that closing the health gap would help reaching the desired social goals. Gender and socioeconomic inequality in health care have received considerable attention in recent literature (Chowdhury and Bhuiya, 1999; Bhuiya et al., 2001) although very few studies have focused on the disparities at the regional perspectives. This study has attempted to look this issue at the wider perspective and demonstrated that the regional differences in the use of health services, in some cases, were greater than the socioeconomic disparities.

Bangladesh continues to face a formidable challenge in the improvement of health of the poor. In a society where incomes of the poor are too low to buy a minimum essential package, the provision should be developed to provide essential health services according to a sliding scale of fees for easily identified subgroups of the population. The health need of the poor should be recognized and health interventions should be tailored to match the specific livelihood strategies of poor households. The distribution of health resources should focus not only on the size of the population but also on the burden of diseases (Whitehead et al., 2001). As a short-term policy measure, targeted health interventions may produce desired outcomes. There are evidences which suggest that targeted approach has the potential to significantly raise access to health services in Bangladesh (Chowdhury and Bhuiya, 1999; Hadi et al., 2001).

Since the focus of the health program should be equitable health development, the current health system should include pro poor health components in it. An essential element of this strategy should be the sensitization of the community about the benefits of this approach, inclusion of the poor in decision making and raising access of the poor to basic health resources and services. As have seen, the health outcomes vary according to socioeconomic categories, the proposed system needs to move beyond the one-size-fits-all model of health care. In other words, the health care for the poor should not only be subsidized but the mode of services must be appropriate to reach them. The policy options to improve maternal health should also include testing new initiatives and systemic interventions that would help designing the most effective intervention models for the poor and disadvantaged. Health development can only be ensured by enhancing the lives of women and by providing them freedom (Sen, 1999). The poor women in Bangladesh should be given that freedom to avoid ill-health during pregnancy and escapable maternal mortality.

The long-term policy options must incorporate several other issues including expansion of health program to the under-served regions, behavioral change issues through adult education among relatively older women and ensuring the availability of trained birth attendants for safe delivery. The region specific inequality may be reduced by the expansion of outreach health programs to bring services closer to the disadvantaged. Among other alternatives, re-allocation of health resources to reduce regional gaps and the promotion of health services for the ethnic minorities and the outreach may be a viable option. The study argues for the development of new approaches which will prioritize the needs of the poorest and most disadvantaged. The reduction of the health inequality can be achieved by adopting targeted health delivery strategy to ensure that the very poor get access to reproductive health services. The study concludes that expanded pro-poor health development program can significantly improve the access to and the utilization of health services among the disadvantaged in developing countries.

Acknowledgements

This project is supported by the Bangladesh Health Equity Watch (BHEW), a collaborative initiative of the Bangladesh Bureau of Statistics (BBS), Bangladesh Institute of Development Studies (BIDS), BRAC and ICCDR, B, funded by the Rockefeller Foundation.

The authors wish to thank three anonymous reviewers for their constructive criticisms and comments of the earlier version of the paper.

References

Baker, J.L. and J. van der Gaag (1993). 'Equity in the finance and delivery of health care: evidence from five developing countries'. In van Doorslaer, A. Wagstaff and F. Rutten (eds). *Equity in the Finance and Delivery of Health Care.* Oxford University Press: London.

Bhuiya, A., M. Chowdhury, F. Ahmed and A. M. Adams (2001). 'Bangladesh: An Intervention Study of Factors Underlying Increasing Equity in Child Survival'. In T. Evans, M. Whitehead, F. Diderichsen, A. Bhuiya, and M. Wirth (eds). *Challenging Inequities in Health. From Ethics to Action.* Oxford University Press: New York.

Chowdhury, A.M.R. and A. Bhuiya (1999). 'Do poverty alleviation programmes reduce inequalities in health? The Bangladesh experience'. In D. Leon and G. Walt (eds). *Poverty, Inequality and Health.* Oxford University Press: Oxford.

Cook, R.J. and M.F. Fathalla (1996). 'Advancing reproductive rights beyond Cairo and Beijing'. International Family Planning Perspectives, 22(3): pp. 115-121.

Evans, T., M. Whitehead, F. Diderichsen, A. Bhuiya and M. Wirth (eds) (2001). *Challenging Inequities in Health. From Ethics to Action.* Oxford University Press, New York.

Feachem, G.A. (2000). 'Poverty and inequity: a proper focus for the new century'. Bulletin of the World Health Organization. 78(1): pp. 1-2.

Filmer, D. and L. Pritchett (2000). 'Estimating wealth effects without expenditure data - tears: An application to educational enrollments in states of India'. *World Bank Policy Research Working Paper No. 1994.* World Bank: Washington DC.

Gwatkin, D., S. Rustein, K. John, R. Pande and A. Wagstaff (2000). *Socio-economic Differences in Health, Nutrition and Population in Bangladesh.* HNP/Poverty Thematic Group of the World Bank. Washington: The World Bank.

Hadi, A., S.R. Nath and A.M.R. Chowdhury (2001). 'The Effects of Micro-credit Programmes on the Reproductive Behaviour of Women in Rural Areas of Bangladesh'. In Z. Sathar and J.F. Philips (eds). *Fertility Transition in South Asia.* Oxford University Press: New York.

Krasnik, A. (1996) 'The concept of equity in health services research'. Scandinavian Journal of Social Medicine. 24(1): pp. 2-7.

Kutzin, J. (2001). 'Obstacles to women's access: Issues and opinions for more effective interventions to improve women's health'. Background paper for the *World Bank Best Practices Paper on Women's Health and Nutrition.* Washington, DC: Population, Health, and Nutrition Department.

Leon, D. and G. Walt (1999). 'Poverty, Inequality, and Health in International Perspective: A Divided World?'. In D. Leon and G. Walt (eds). *Poverty, Inequality and Health.* Oxford University Press: Oxford.

Magadi, M.A., N.J. Madise and R.N. Rodrigues (2000). 'Frequency and timing of antenatal care in Kenya: explaining the variations between women of different communities'. Social Science and Medicine. 51: pp. 551-561.

Makinen, M., H. Water, M. Rauch, et al. (2000). 'Inequalities in health care use and expenditures: empirical data from eight developing countries and countries in transition'. Bulletin of the World Health Organization. 78(1): pp. 55-65.

Miles-Doan, R. and K.L. Brewster (1998). 'The impact of the type of employment, use of prenatal services and contraceptive practice in the Phillipines'. Studies in Family Planning. 29: pp. 69-77.

Mitra, S.N., A. Al-Sabir, A.R. Cross and K. Jamil (1997). *Bangladesh Demographic and Health Survey 1997-1997.* NIPORT, Mitra Associates and Macro International: Dhaka.

Mitra, S.N., Al-Sabir, A., T. Saha and S. Kumar (2001). *Bangladesh Demographic and Health Survey 1999-2000.* NIPORT, Mitra Associates and Macro International: Dhaka.

Power, J.G. (1996). 'Evaluating health knowledge: an alternative approach'. Journal of Health Communication. 1: pp. 285-298.

Prasad, P. (2000). 'Health care access and marginalized social spaces. Leptospirosis in South Gujarat'. Economic and Political Weekly. 35(41): pp. 3688-3694.

Sen, A. (1999). 'Health in development'. Bulletin of the World Health Organization. 77(8): pp. 619-623.

Swenson, I.E., N.M. Thang, V.Q. Nhan and P.X. Tieu (1993). 'Factors related to the utilization of prenatal care in Vietnam'. Journal of Tropical Medicine and Hygiene. 96: pp. 76-85.

Wagstaff, A. (2002). 'Poverty and health sector inequalities'. Bulletin of the World Health Organization. 80(2): pp. 97-105.

Whitehead, M., G. Dahlgren, and L. Gilson (2001). 'Developing the Policy Response to Inequities in Health: A Global Perspective'. In T. Evans, M. Whitehead, F. Diderichsen, A. Bhuiya and M. Wirth (eds). *Challenging Inequities in Health. From Ethics to Action*, Oxford University Press: New York.

WHO (2000). 'Pregnancy Exposes Women in poor states to 200-fold risk of death, compared with rich ones, says WHO'. Populi. 27(2). pp. 4.

World Bank (1993). *World Development Report 1993. Investing in Health*. Washington DC: Oxford University Press.

Chapter 11

Development, Validation and Performance of a Rapid Consumption Expenditure Proxy for Measuring Income Poverty in Tanzania: Experience from AMMP Demographic Surveillance Sites

Philip Setel, Savitri Abeyasekera, Patrick Ward, Yusuf Hemed, David Whiting, Robert Mswia and Manos Antoninis

Summary

This chapter describes the production and validation of a rapid consumption expenditure proxy (CEP) tool for measuring income poverty at the household level. The work was undertaken through the Adult Morbidity and Mortality Project, Phase 2 (AMMP-2) in three demographic surveillance sites. The rapid poverty measurement tool described here can also be used as an add-on to smaller research projects or monitoring activities with a predominantly qualitative or participatory approach. In this way qualitative/participatory and quantitative information gathering strategies can be used to augment each other. For our purposes a suitable poverty measure had to be able to generate data comparable to other national sources and be able to provide indicators useful to those outside the health sector.

We first developed an initial model using preliminary Household Budget Survey data provided by the Tanzania Bureau of Statistics, followed by fieldwork and data collection. We then undertook validation of the CEP model using variables collected by AMMP, and the complete Household Budget Survey data set. Separate CEP models were developed for two rural DSS sites and for one urban DSS site. The dependent variable in the model was the logarithm of consumption expenditure. The candidate explanatory variables were then dropped one-by-one from the CEP regression model to see if their exclusion had a significant effect on the final prediction of expenditure. The squared differences in predicted and observed expenditure for all of the individual households were summed to assess model performance.

In all three regions, the mean of actual values and the mean of predicted values were very close. The percent error in the predictions was also quite small and below 17%, but this must be treated with some caution since it relates to log values. About 60% of the variability in the consumption expenditure is explained by these models.

We believe the CEP yields results that can be used with confidence in the context of DSS or other survey work to monitor the experience of broad income-poverty groups in Tanzania with respect to important indicators of health, well being, and survival. The models are at their best when predicting values around the mean. A CEP approach deserves careful consideration for research and analysis in countries where appropriate data exist to develop and model the proxies.

Background

Attacking poverty and its connections with health has become a major global priority [1, 2]. High profile programs such as the Highly Indebted Poor Countries/Poverty Reduction Strategy (PRSP) initiative of multi-lateral debt forgiveness, and the Millennium Development Goals require the development of evaluable national poverty reduction strategies and quantifiable indicators for measuring progress against those goals. It has become increasingly apparent that targets expressed simply as 'national averages' are inadequate to ensure that the poorest citizens of poor countries benefit from poverty reduction measures. Unless specific safeguards are taken to target and track conditions among the poorest of the poor, it is possible that countries may achieve their targets without significantly improving conditions among them. In order to assess whether the needs of the poorest are being met, a sound method for identifying them is required.

The need for poverty monitoring and impact data could not be greater. Yet there is still a great deal of debate and exploration about 'best practice' with respect to rapid, reliable, validated, and feasible tools to measure poverty and health equity in developing countries[3-6].

Here we describe the production and validation of a rapid consumption expenditure proxy (CEP) tool for measuring income poverty at the household level. The work was undertaken through the Adult Morbidity and Mortality Project, Phase 2 (AMMP-2). AMMP is a project of the Tanzanian Ministry of Health and District/Municipal councils, funded by the UK Department for International Development, and implemented in partnership with the University of Newcastle upon Tyne.

One of AMMP's main activities is to support the Ministry of Health and local councils to operate several demographic surveillance sites (DSS). Working through the Ministry of Health and local councils, AMMP has operated one urban and two rural DSS sites since 1992. These sites are located in the Temeke and Ilala Municipalities in Dar es Salaam Region (14,000 households), Hai District in Kilimanjaro Region (31,000 households), and Morogoro District in Morogoro Region (29,000 households). They were originally selected because they were geographically dispersed and, according to data available at the time, thought to represent a range of urban and rural living standards. In 2002-2003, two additional sites were established in Igunga (Tabora Region) and Kigoma Urban/Ujiji (Kigoma Region).[1] The methods and outputs of the AMMP DSS have been described in detail elsewhere[7-9]. In brief, demographic surveillance under AMMP consists of:

---

1   As of this writing, data from Igunga and Kigoma Urban were not available for analysis.

- annual updates of all demographic events (births, deaths, and in- and out-migrations) within a geographically defined population at each DSS site;
- continuous cause-specific mortality monitoring using verbal autopsy techniques; and
- periodic implementation of comprehensive or sample surveys of morbidity, risk factors or other characteristics.[2]

One of the chief aims of AMMP is to use DSS to provide accurate community-based evidence about the major causes of ill health and death among the poorest members of Tanzanian society, and to contribute to equitable development in the country.

It is hoped that the approach we have taken to measuring income poverty in Tanzania may be of use to those engaged in the development of similar tools elsewhere. In particular, it may be of use in generating data on poverty and equity using DSS, sample vital registration or large sample surveys. The rapid poverty measurement tool described here can also be used as an add-on to smaller research projects or monitoring activities with a predominantly qualitative or participatory approach. In this way qualitative/participatory and quantitative information gathering strategies can be used to augment each other.

It is important to note that replicating the methods described here requires access to data from a recent household budget or living standards measurement survey, or a similar large-sample household survey with a measure of consumption expenditure as an output. Even if this is lacking, some of the principles of the development process we used may still be useful.[3]

Defining an Approach

Theoretical constructs of poverty abound, as do methods for their measurement [*10*]. In order to define an approach, we first considered the general context of information need, and then identified the poverty construct best suited to it. We then reviewed the literature on options for measuring that construct, and lastly considered any overriding practical constraints to measuring the desired poverty construct in a

---

2   For a general discussion of demographic surveillance systems in Africa and in Asia see INDEPTH Network, Editor. *Population and Health in Developing Countries. Vol 1. Population, Health and Survival at INDEPTH Sites*. Ottawa: International Development Research Centre; (2002).

3   The full text of the consultant reports describing these methods (Antoninis, M., 'Socio-economic Status Predictors for the Adult Morbidity and Mortality Project Census in the Hai and Morogoro Rural Districts'. (2000), Adult Morbidity and Mortality Project, Tanzanian Ministry of Health: Dar es Salaam; Antoninis, M., 'Socio-economic Status Predictors for the Adult Morbidity and Mortality Project Census in the Ilala and Temeke Districts of Dar es Salaam'. (2000), Adult Morbidity and Mortality Project, Tanzanian Ministry of Health: Dar es Salaam; and Abeyasekera, S. and P. Ward, 'Models for Predicting Expenditure per Adult Equivalent for AMMP sentinel surveillance sites'. (2002), Adult Morbidity and Mortality Project, Tanzanian Ministry of Health: Dar es Salaam) can be found on the AMMP website at http://www.ncl.ac.uk/ammp.

technically sound manner. These constraints included considerations such as ease of implementation and integration into the DSS operations, resource needs, and existing in-country experience. For our purposes a suitable poverty measure had to be:

- rapid and inexpensive to implement;
- easily integrated into the routine annual and semi-annual census update rounds;
- be able to identify households with respect to national poverty lines;
- be able to generate data comparable to other national sources; and
- be able to provide indicators useful to those outside the health sector.

Because of the need to be able to produce absolute measures of income poverty, we elected not to use an approach based on asset lists or asset indices, as these can produce only relative measures and are seldom validated for predictive accuracy.

The Context of Information Need

AMMP is rooted in the Health Sector and in health sector reform. Three core principals of health reform in Tanzania are:

- supporting evidence-based policy and planning at the district and national levels;
- supporting the decentralisation process in which districts are primarily responsible for delivering care and for planning, and the central ministry is responsible for steering policy and setting guidelines for practice; and
- improving health system accountability to locally served populations and enhancing equity and service to the poorest and most vulnerable members of society.

Health reform, however, is itself embedded in wider programs of national reform and in multi-sectoral initiatives. For example, at the district level health reform is subsumed within a country-wide process of government reform based on devolution to local council authority that has been in train since the mid 1990s. Local government reform is being steered by the President's Office. More recently, the country has committed itself to meeting and monitoring progress toward targets set in the Poverty Reduction Strategy Paper and Poverty Monitoring Master Plans. [*11, 12*] The PRSP process and the monitoring and evaluation effort are located in the Vice President's Office. In addition, Tanzania's renewed commitment to fight the AIDS epidemic was made in 2000 with the establishment of the Tanzanian AIDS Commission in the Prime Minister's Office. These national reforms and initiatives, based as they are at high levels of Ministerial, Presidential, Vice-Presidential, and Prime-Ministerial office each have urgent needs for poverty, demographic, and health data.

The most cogent statement about how all these needs are to be met in a resource-poor setting like Tanzania comes from the Poverty Monitoring Master Plan: 'Tanzania is moving towards a co-ordinated national-level approach to data and information collection, analysis, and dissemination ... [and] ... away from single-purpose information generation toward a multi-purpose and interlinked approach at the national level' (p. 3).

The PRSP also contains the most explicit definitions of what is meant by 'poverty' and who are 'the poor.' Other programs are less clear. The two constructs most germane to the poverty reduction spelled out in the PRSP are 'income' and 'non-income' poverty [13]. The former is defined in the PRSP principally in terms of poverty lines, such as those who live on one US dollar per day. The PRSP defines the latter as the degree of access to social and community services like schools and a safe water supply.

The project elected first to focus on the measurement of income poverty. It was felt that too many fundamental theoretical issues about the use and interpretation of measures of non-income poverty constructs such as 'social capital' remain unresolved [14, 15].

Consumption expenditure per adult equivalent is possibly the most appropriate measure of long-term or 'permanent' income for developing countries [4, 16]. Consumption expenditure is also one of the principle outcome variables in national household budget (or similar) surveys and forms the basis for the calculation of poverty estimates in most developing countries [17]. The measurement of consumption expenditure in these surveys requires an intensive set of methods that include the keeping of detailed household expense diaries and the repeated administration of lengthy survey questionnaires.

It would not have been practical to consider replicating the full survey methods necessary for such a study within the context of a technically simple DSS administered by district-level health staff, particularly if measures were wanted from every household. Therefore, we elected to pursue the general approach of a rapid validated measure of socio-economic status proposed by Morris et al. [3]. We also developed collaborative links with the National Bureau of Statistics, who administered the National Household Budget Survey [17].

Our objective was to develop a tool to derive a validated consumption expenditure proxy measure, or what we have termed a 'CEP' tool. The advantage of this approach was that it would allow us to predict household consumption levels using information from a limited set of questions that could be easily and cheaply collected from all households during project census update rounds. For the sake of simplicity we will refer to this measurement as 'consumption expenditure' or simply 'expenditure.'

Methods

Development of the CEP tool for measuring income poverty at the household level was carried out in two phases. First came initial model development using preliminary Household Budget Survey data provided by the Bureau of Statistics, followed by fieldwork and data collection using the preliminary models. Second, we undertook validation of the CEP model using variables collected by AMMP, and the complete Household Budget Survey data set from the regions in which AMMP DSS sites are located. The second phase also included a process of CEP model minimisation (i.e. elimination of variables found not to improve the estimates of consumption expenditure), and tests of the performance of the final model. This reliability testing included measures of how well the final model predicted true household consumption expenditure, how accurately it allocated households to expenditure terciles and quintiles, and how well it predicted the status and proportions of poor and non-poor households *vis a vis* the basic needs poverty line.

As mentioned, the approach we have taken requires access to data from a recent national household budget survey or living standards measurement-type survey, and preferably to data from sampling clusters in or near the geographic location of the DSS site or sites for which the proxy is to be developed. In addition, the input of a qualified statistician is required to replicate the approach described here. This input is needed to produce and test the CEP models with appropriate adaptations to local conditions and data sources. Once the CEP data have been collected in the field, estimates of household consumption expenditure can be derived.

Preliminary CEP Model Development and Data Collection

Separate CEP models were developed for the two rural DSS sites and for the urban DSS site using a preliminary data set of Household Budget Survey made available through the collaboration with the Bureau of Statistics [18, 19].[4] For the rural areas, we were provided with the first available data of the 2000-2001 Household Budget Survey collected from May to August 2000. The sample size of 1,308 households in the preliminary data set was a small fraction of the 22,000 households that were eventually contacted during the twelve months of fieldwork for the national survey (details of the Household Budget Survey methods are contained in the survey report). However, these were the only data available at the time AMMP required work to begin on the development of the poverty measurement tool [18, p. 3]. For the urban model, data on only 280 households collected between May and July 2000 were available [19, p. 4].

Potential questions and variables to be included in the CEP were drawn from the entire range of data collected on the Household Budget Survey.[5] In addition, we included variables collected in the Household Budget Survey that are intrinsic to health (e.g. a household's water source), and specific commodities or assets felt to be related to wealth status in the different DSS sites (e.g. ownership of livestock such as cattle). In designing the CEP survey tool using these variables it was important to frame questions to yield identical or nearly-identical response categories to those contained in the Household Budget Survey.

The preliminary CEP models contained approximately 50 items each, most of which required single-response or yes/no response categories. The candidate predictors included variables such as housing materials, consumption of meat in the last week, and expenditures on fertilizer.

In the 2000 AMMP census update rounds data were collected from 13,223 households in the urban NSS/AMMP DSS area, and 59,755 households in the two

---

4 This preliminary step of CEP model development was necessary because the CEP tool was required for the 2000 round of AMMP census updates. The full Household Budget Survey data set was not ready for analysis in time for the development of the AMMP CEP tool.

5 Candidate variables were identified using a stepwise regression procedure that modelled expenditure on a wide range of potential poverty proxy variables. This modelling involved specifying a selection probability level of 0.2 and highlighting predictors significant at the 5% and 1% levels in the presentation of results.

rural sites. After the model minimisation process using the full Household Budget Survey data set (discussed below), these data were used to generate the main outcome of interest: an estimate of monthly household consumption expenditure per adult equivalent. The minimised models were then slightly modified in light of a national proxy model development effort [20] and used in 2002 to collect data from approximately 37,000 households in two additional DSS sites, one urban (Kigoma Urban, Kigoma Region) one rural (Igunga District, Tabora Region).

Analytical Method for CEP Model Finalisation and Validation

The objective of the second phase of work was to take the larger preliminary models and further reduce them to the set of variables that best predicted consumption expenditure. The dependent variable in the model was the logarithm of consumption expenditure that was transformed after model development into a monetary value in Tanzanian Shillings (TSh).[6] The final minimal models were developed using the full set of data available from the Household Budget Survey, although analysis was restricted to proxy variables that were collected for the AMMP households.

Selection of explanatory variables for inclusion in the model was the first step in the final CEP model development. A variety of factors were taken into account in doing so. These included: statistical consideration such as discriminatory power, non-statistical factors such as the intrinsic importance of particular variables, and pragmatic concerns such as ease and reliability of repeated measurement. In terms of statistical selection criteria, the preliminary models relied heavily on the 'coefficient of determination' (R^2). This measures the proportion of variation in expenditure that is explained by the set of predictor variables.[7] In other words, it indicates how well the variation in expenditure is accounted for by the models – including the candidate predictor variables.

AMMP data were available for the DSS sites in Temeke/Ilala (Dar es Salaam) and in Hai and Morogoro districts. As noted above, these are located in the Dar es Salaam, Kilimanjaro and Morogoro regions, respectively. Therefore, data from the Household Budget Survey for these three regions were separately modelled using the logarithm of consumption expenditure as the response variable and the proxy variables identified in the previous phase of work as potential explanatory variables. After a series of iterations involving fitting and evaluating models with different groups of predictors (i.e. variables in common from the Household Budget Survey data and the preliminary CEP data) the most suitable models for predicting expenditure were determined for each region.

To start with, all the variables from the preliminary AMMP CEP models were included from the data on the Household Budget Survey households. The candidate variables were then dropped one-by-one from the CEP regression model to see if their

---

6   The regression approach used in model development requires the response being modelled to have a normal distribution. Consumption expenditure has a very skew distribution and transforming to logarithms provided a normally distributed variable.

7   While high values of R^2 are desirable and provide a quick summary of how well the data fit the model, reliance on R^2 alone has limitations. For example, it cannot discern patterns in the relationship showing deviations from linearity.

exclusion had a significant effect on the final prediction of expenditure. If removing a variable substantially changed the expenditure estimate, that variable was left in the final CEP model.[8] The residuals[9] corresponding to this model were then plotted to check model assumptions.

The first step in the validation process involved plotting actual values against predictions to give an overall assessment of how well the model can be used for predictive purposes. We did this for a random selection of observations from the Household Budget Survey that were held out of the model building process. Using a different set of observations in this way is important because applying the CEP model to the same set of observations that generated it will, of course, be expected to give better results than if the CEP were applied to the 'hold back' sample mentioned above. We then reversed this process, using the original hold-back sample to determine the final CEP model and validating it on the original data.

Another useful method that provides a comparison between alternative models involves fitting the selected model N times (N = number of households in the data set) with $N-1$ cases, each time omitting one household in turn and predicting its expenditure from the model fitted without its inclusion. The squared differences in predicted and observed expenditure for all of the individual households are then summed to assess model performance.[10]

Results

Model Results

Predictor variables that were found to be common across the three regions were as follows:

- household size
- education level of head of household in 4 classes (none; primary; secondary; tertiary and above)
- number of days meat eaten in past week

---

8   Variables eliminated at a previous stage were then re-considered in turn to confirm if they were still non-significant in the model.

9   The differences between observed and predicted values are called 'residuals.' A plot of residuals against the predicted values checks the model for validity of model assumptions (variance homogeneity, and normality). This will also highlight any observations that do not fit the model well.

10 This is called the Predicted Residual Sum of Squares, or PRESS statistic. Fortunately, the PRESS statistic does not require fitting the model N times. Theoretically it can be shown that the PRESS statistic can be calculated using the simple residuals from a single fit of the model to all the data and the *leverages* of each data point. The latter are a measure of influence of the predictors on the predicted value. If h_i is the leverage of the ith observation, then the PRESS statistic is given by PRESS = Σ [(observed y_i − predicted y_i) ÷ (1 − h_i)]^2 (see [21]). PRESS statistics are easily calculated from a single model fit using results generated in standard statistics software packages such as SPSS and STATA.

Other predictor variables included in the models for these three regions were as follows:

Dar es Salaam (urban):

- whether household owned an iron, an electric/gas stove, an automobile
- construction materials of walls in 2 classes (modern; not modern)
- number of days in past week when milk products were consumed
- in past month whether household paid money to purchase wheat flour, cooking bananas, potatoes, fresh fish, beer, newspapers, poultry products, eggs, a snack or beverage outside household.

Kilimanjaro (rural):

- age of household head
- area of land used for farming/pastoralism
- whether household spent money to purchase seeds in the past 12 months
- whether household spent money to purchase fertiliser or manure in the past 12 months
- whether household owned a bicycle, a sofa, a lamp
- main source of cash income in the following 4 classes (sale of cash/food crops; sale of livestock or livestock products; business/wages/salaries; other sources [e.g. fishing, casual cash earnings, cash remittances])

Morogoro (rural):

- sex of household head
- age (in years) of household head
- number of persons employed in household (including self employed)
- dependency ratio (number of dependants ÷ number of non-dependants)
- number of persons per sleeping room
- whether household spent money to purchase fertiliser or manure in the past 12 months
- main source of drinking water in 4 classes (piped in house; piped outside house; protected; unprotected)
- construction materials of walls in 2 classes (modern; not modern)
- toilet facility available (none; not modern; modern)
- main fuel used for lighting (electricity; kerosene/paraffin; other)
- whether household owned a bicycle, a bed net

The adjusted R^2 values for the above models were 65% for Kilimanjaro region, 56% for Morogoro region and 63% for Dar es Salaam region.

Model Validation

Summary statistics that provide an assessment of the validity of the fitted models, when applied to a test data set, are shown for each region in Table 11.1. Each row

Table 11.1   Summary statistics of true values and predictions for households remaining in subset B, from results of model fitted to data of a random subset A for Dar es Salaam, Kilimanjaro, and Morogoro Regions

	Mean	Std. Dev	Min	Max
Dar es Salaam region (N=611)				
y = log$_e$ (expenditure)	9.65	0.644	6.18	11.59
y prediction	9.73	0.517	8.14	11.02
% error in prediction	3.46	2.76	0.0011	16.47
true minus predicted	−0.0316	0.404	−1.31	1.29
Kilimanjaro region (N=511)				
y = log$_e$ (expenditure)	9.44	0.617	7.75	11.60
y prediction	9.47	0.470	8.48	11.19
% error in prediction	3.44	2.94	0.00013	15.91
true minus predicted	−0.0325	0.420	−1.343	1.146
Morogoro region(N=526)				
y = log$_e$ (expenditure)	9.51	0.614	7.44	11.86
y prediction	9.48	0.506	8.25	10.88
% error in prediction	3.56	2.86	0.0082	16.89
true minus predicted	0.0350	0.432	−1.430	1.633

represents a summary across individual household values. The percent error in prediction has been calculated as:

$$\{|true - predicted| \div true\} \times 100$$

In all three regions, the mean of actual values and the mean of predicted values are very close. There is also a close correspondence between actual and predicted values at the household level, as is observed in Figure 11.1 for Morogoro region (below). The measure of linear association between the predicted and true expenditure values (expressed by the correlation coefficient 'r') was 0.72. This indicates a relatively high degree of association. Similar graphs were obtained for Kilimanjaro and Dar es Salaam regions.

The percent error in the predictions is also quite small and below 17%, but this must be treated with some caution since it relates to log values. Reverse transformation into TSh is simpler to interpret using the actual differences between predicted and true expenditure (shown in the last row of Table 11.1). The worst scenario is in (Morogoro). When converted to TSh the maximum of the differences between true and predicted values (1.633) gives a value 5.1. This means that at worst, there is a five-fold error in the predictions.[11] However, the mean difference, converts to a value

---

11 The results re-transformed in TSh must be interpreted in a multiplicative fashion since a difference in logs is the same as the log of a ratio, i.e. log A − log B = log (A ÷ B).

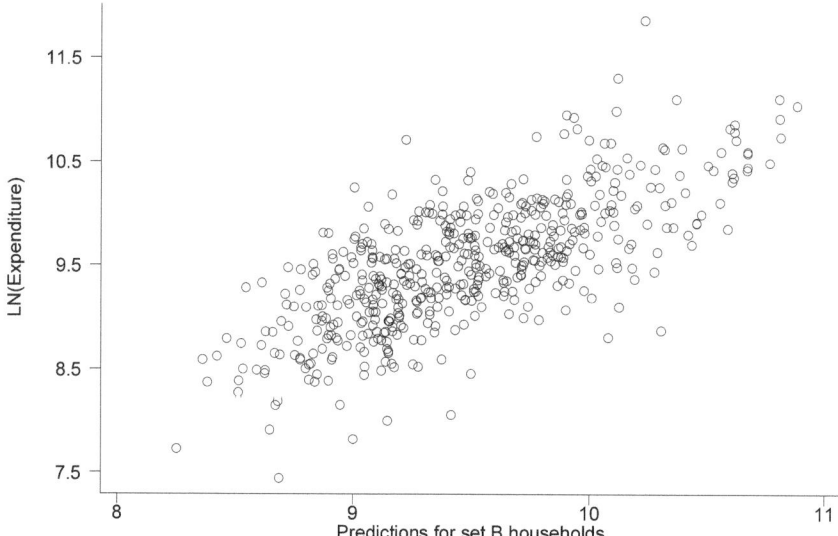

Figure 11.1 True values vs. predictions for (remaining) subset B of households using model fitted to a random subset (A) of households (Morogoro region, N=526; r = 0.72)

of 1.036, i.e. virtually a 1:1 correspondence in means between the true and predicted values.

Model Performance

In Table 11.2 we present results intended to assess how well the CEP performs. The purpose of this performance assessment is to aid in the interpretation of results from the CEP and to allow a general assessment of whether it can serve policy and/or research purposes. The first two table columns show the actual percentage of the population below the poverty line and the corresponding percents as predicted from the model. Because it uses far fewer variables in predicting expenditure than does the Household Budget Survey, the CEP predictions tended to cluster or compress toward the mean. This led to a substantial under-estimate in the initial prediction of the percent of households below the poverty line. Applying the appropriate correction procedure developed by Hentschel et al. [22] produced much better estimates (third numerical column in Table 11.2). The final two columns of the table give the percentage of households correctly classified by the poverty line, and the percentage classified into the correct tercile. These results are as good as can be expected, given that only about 60% of the variability in the consumption expenditure is explained by these models.

Table 11.2   Indicators of model performance

Region	Actual % below poverty line	Predicted % below poverty line	% below poverty line applying Hentschel approach after prediction	% correctly classified by poverty line	% classified into the correct tercile
Dar es Salaam	11.0	4.9	11.4	90.9	62.8
Kilimanjaro/ Hai	18.9	13.5	18.1	87.0	60.7
Morogoro	21.3	12.8	22.9	83.9	60.3

Discussion and Recommendations

The AMMP CEP tool resulted from an explicit process that had several objectives and worked within several constraints. Chief among these were the need to address the indicator requirements of a multi-sectoral audience, and to provide a technically sound poverty measure with results that could be compared to other data sets. This led us to model consumption expenditure and to link our efforts to the analysis of the National Household Budget Survey.

We believe the CEP yields results that can be used with confidence in the context of DSS or other survey work to monitor the experience of broad income-poverty groups in Tanzania with respect to important indicators of health, well being, and survival. It is important to note that while we used a CEP approach to build models specific to each region in which AMMP DSS is located, the technique is also applicable in a policy context at a national level. Ward et al. [20] used nearly identical methods to derive a set of proxy predictors for household expenditure and income poverty meant to be incorporated into sample surveys undertaken in Tanzania.

The regional CEP models performed reasonably well with respect to predictions of mean expenditure estimates for groups of households, and can therefore be reliably applied when model-based consumption expenditure values are averaged to the community level. Use of these model predictions that do not involve any averaging process but rely on individual household level consumption expenditure figures must be regarded with some caution, since misclassification errors were quite high for individual households close to the boundary of the poverty line.

The regional CEPs predicted log values of consumption expenditure extremely well. The models are at their best when predicting values around the mean. At their worst, when transformed into 'real' values of TSh, one of our models (Morogoro) had up to a five-fold error in the prediction of the correct value. Because these models are intended to identify and chart the experiences of the poor, it will be important to look at ways of improving model accuracy for the lower end of the expenditure distribution. Nevertheless, the CEP did correctly identify the poorest third of the population with about 60% accuracy and the poorest 20% of the population with over 50% accuracy. In addition, misclassification of households into non-neighbouring quintiles was extremely low.

As analysis contained in separate reports and presentations demonstrates [*19, 23, 24*], this approach, where it can be applied, is a general improvement over more commonly used asset indices and lists [*25, 26*]. A rapid CEP tool for measuring income poverty has several features that may be of use and interest to those engaged in similar work in other developing countries. In Tanzania, the general move toward multi-purpose information generation makes maximal and rational use of scarce resources for poverty reduction and health sector monitoring, and should be emulated. This is especially true of countries where the health sector will be expected to produce equity- and poverty-sensitive indicators, and those outside the health sector are expected to produce indicators about major health-related issues. Within such a context, the development of a CEP-type of rapid measurement tool within the health sector can be used as a vehicle to bring Ministry of Health and National Bureaus of Statistics into closer collaboration. Despite their mutual areas of interest, there is sometimes little contact and cross-development of capacity at a country level.

A CEP approach deserves careful consideration for countries where the data exist to generate it. Capacity can be put in place to allow the process to be replicated following subsequent Household Budget Surveys. We also believe that the apparent degree of increased accuracy in comparison with other options, together with the ability to validate the CEP against a 'gold standard' and to derive an absolute (and not just a relative) poverty measure justifies the additional input of resources over re-analysis or re-application of existing asset indices based on asset lists from DHS surveys.

The primary goal of the CEP is to serve social policy and broad impact evaluation needs. We believe that it can meet that objective while offering a potentially higher standard of performance even for research applications.

Acknowledgements

This publication is, in part, an output of the Adult Morbidity and Mortality Project (AMMP). AMMP is a project of the Tanzanian Ministry of Health, funded by the Department for International Development (UK), and implemented in partnership with the University of Newcastle upon Tyne (UK). The views expressed are not necessarily those of DFID.

The AMMP Team includes: KGMM Alberti, Richard Amaro, Yusuf Hemed, Gregory Kabadi, Berlina Job, Judith Kahama, Joel Kalula, Ayoub Kibao, John Kissima, Henry Kitange, Regina Kutaga, Mary Lewanga, Frederic Macha, Haroun Machibya, Mkamba Mashombo, Godwill Massawe, Gabriel Masuki, Louisa Masayanyika, Ali Mhina, Veronica Mkusa, Ades Moshy, Hamisi Mponezya, Robert Mswia, Deo Mtasiwa, Ferdinand Mugusi, Samuel Ngatunga, Mkay Nguluma, Peter Nkulila, Seif Rashid, JJ Rubona, Asha Sankole, Philip Setel, Nigel Unwin, and David Whiting.

References

1. Evans, T., M. Whitehead, F. Diderichsen, A. Bhuiya, and M. Wirth, eds. *Challenging Inequities in Health. From Ethics to Action.* 2001, Oxford University Press: New York.
2. Jha, P. and A. Mills, *Improving Health Outcomes of the Poor. Report of Working Group 5 of the Commission on Macroeconomics and Health.* (2002), World Health Organization: Geneva.
3. Morris, S.S., C. Carletto, J. Hoddinott, and L.J.M. Christiaensen, 'Validity of rapid estimates of household wealth and income for health surveys in rural Africa'. Journal of Epidemiology and Community Health, (2000). 54: pp. 381-387.
4. Montgomery, M.R., M. Gragnolati, K.A. Burke, and E. Paredes, 'Measuring Living Standards with Proxy Variables'. Demography, (2000). 37(2): pp. 155-174.
5. Anand, S., F. Diderichsen, T. Evans, V.M. Shkolnikov, and M. Wirth, *Measuring Disparities in Health: Methods and Indicators*, in *Challenging Inequities in Health. From Ethics to Action*, T. Evans, M. Whitehead, F. Diderichsen, et al., Editors. 2001, Oxford University Press: New York. pp. 49-67.
6. Leon, D. and G. Walt, 'Poverty, inequality, and health in international perspective: a divided world?', in *Poverty, Inequality, and Health. An International Perspective*, D. Leon and G. Walt, Editors. (2001), Oxford University Press: Oxford. pp. 1-16.
7. Mswia, R., D. Whiting, G. Kabadi, H. Masanja, P. Setel, and for the AMMP Team, *Dar es Salaam Demographic Surveillance System*, in *Population and Health in Developing Countries. Volume 1: Population, Health, and Survival in INDEPTH Sites*, The INDEPTH Network, Editor. (2002), International Development Research Centre: Ottawa. pp. 143-150.
8. Mswia, R., D. Whiting, G. Kabadi, H. Masanja, P. Setel, and for the AMMP Team, *Hai District Demographic Surveillance System*, in *Population and Health in Developing Countries. Volume 1: Population, Health, and Survival in INDEPTH Sites*, The INDEPTH Network, Editor. (2002), International Development Research Centre: Ottawa. pp. 151-158.
9. Mswia, R., D. Whiting, G. Kabadi, H. Masanja, P. Setel, and for the AMMP Team, *Morogoro Rural Demographic Surveillance System*, in *Population and Health in Developing Countries. Volume 1: Population, Health, and Survival in INDEPTH Sites*, The INDEPTH Network, Editor. (2002), International Development Research Centre: Ottawa. pp. 165-172.
10. Lynch, J. and G. Kaplan, 'Socioeconomic Position', in *Social Epidemiology*, L.F. Berkman and I. Kawachi, Editors. (2000), Oxford University Press: Oxford. pp. 13-35.
11. United Republic of Tanzania, *Poverty Reduction Strategy Paper (PRSP).* (2000): Dar es Salaam (Government Printers).
12. United Republic of Tanzania, *Poverty Monitoring Master Plan.* (2001), Dar es Salaam: Government Printers.
13. Hakikazi Catalyst, *Tanzania without Poverty. A Plain Language Guide to Tanzania's Poverty Reduction Strategy Paper.* (2001), Arusha (Tanzania): Hakikazi Catalyst.
14. Astone, N.M., C.A. Nathanson, R. Schoen, and Y.J. Kim, 'Family Demography, Social Theory and Investment in Social Capital'. Population and Development Review, (1999). 25(1): pp. 1-31.
15. Kunitz, S.J., 'Accounts of Social Capital: The Mixed Health Effects of Personal Communities and Voluntary Groups', in *Poverty, Inequality, and Health. An International Perspective*, D. Leon and G. Walt, Editors. (2001), Oxford University Press: Oxford. pp. 159-174.

16. Montgomery, M.R., M. Gragnolati, K. Burke, and E. Paredes, 'Measuring Living Standards with Proxy Variables'. Working Paper No 129. (1999), The Population Council: New York.
17. National Bureau of Statistics Tanzania, *Household Budget Survey 2000/01.* (2002), National Bureau of Statistics: Dar es Salaam.
18. Antoninis, M., *Socio-economic Status Predictors for the Adult Morbidity and Mortality Project Census in the Hai and Morogoro Rural Districts.* (2000), Adult Morbidity and Mortality Project, Tanzanian Ministry of Health: Dar es Salaam.
19. Antoninis, M., *Socio-economic Status Predictors for the Adult Morbidity and Mortality Project Census in the Ilala and Temeke Districts of Dar es Salaam.* (2000), Adult Morbidity and Mortality Project, Tanzanian Ministry of Health: Dar es Salaam.
20. Ward, P., T. Owens, and G. Kahyrara, *Developing Proxy Predictors for Household Expenditure and Income Poverty.* (2002), Oxford Policy Management: Oxford.
21. Montgomery, D.C. and E.A. Peck, *Introduction to Linear Regression Analysis.* (1982), New York: Wiley.
22. Hentschel, J., J.O. Lanjouw, P. Lanjouw, and J. Poggi, 'Combining Census and Survey Data to Study Spatial Dimensions of Poverty: A Case Study of Ecuador'. The World Bank Economic Review, (2000). 14(1): p. 147-165.
23. Abeyasekera, S. and P. Ward, *Models for Predicting Expenditure per Adult Equivalent for AMMP Sentinel Surveillance Sites.* (2002), Adult Morbidity and Mortality Project, Tanzanian Ministry of Health: Dar es Salaam.
24. Setel, P., S. Abeyasekera, P. Ward, Y. Hemed, D. Whiting, R. Mswia, M. Antoninis, H. Kitange, and for the Adult Morbidity and Mortality Project and National Sentinel Surveillance System Teams, 'Development, Validation, and Performance of a Rapid Consumption Expenditure Proxy for Measuring Income Poverty in Tanzania: Experience from AMMP Demographic Surveillance Sites'. *Paper presented at DFID Asset Index Seminar, Slough, UK, March 21-23.* (2003), Adult Morbidity and Mortality Project, Tanzanian Ministry of Health: Dar es Salaam.
25. DFID Health Systems Resource Centre, *DFID Asset Index Seminar Report. Slough (UK).* (2003), DFID Health Systems Resource Centre: London.
26. Diamond, I., Z. Matthews, and R. Stephenson, *Assessing the health of the poor: Towards a pro poor measurement strategy.* (2001), The Health Systems Resource Center: London.

Chapter 12

Assessing Economic Inequalities in Health: Contributions of the INDEPTH Health Equity Project

Davidson R. Gwatkin

Introduction

In recent years, a long-standing concern about health inequalities between poor and rich countries has been joined by increasing attention to inequalities between poor and better-off groups within countries. The INDEPTH health equity project has taken movement in this direction a step further, by examining economic inequalities in health among groups in much smaller areas.

In doing so, the project investigators have contributed to the assessment of health inequalities in two ways:

- First, by demonstrating the frequent presence of significant economic inequalities in health status and health service use within in relatively small, seemingly homogeneous geographic areas.
- Second, by helping confirm the feasibility of including the study of economic inequalities through health surveys.

Contribution One: Demonstrating the Frequent Presence of Significant Inequalities within Small Areas

The Presence of Small-Area Inequalities

A priori, it might seem reasonable to expect any disparities found within a small geographic area to be modest, especially compared with those reported for the country in which the area is located, because of the much greater social and economic diversity normally found in the country's far larger area and population. However, the findings presented in this volume suggest that such is by no means always the case, at least with respect to equity's economic dimension.[1] Significant economic

---

1   Equity is a multi-dimensional concept that includes inequalities not only in economic status, but also in such other important aspects of well-being as gender, religion, nationality, ethnicity, language, and occupation. In focusing primarily on the economic

inequalities in health status and health service use appear within a clear majority of the small sites covered by the INDEPTH equity studies.

Two annexes provide summary information on the seven sites whose work permits comparisons across economically-defined groups of people.[2] Annex 12.1 deals with differences in health status; Annex 12.2 covers health service use.

Inequalities in Health Status

Frequency and magnitude   The principal health status indicators covered in the INDEPTH studies were infant and under-five mortality. These were approximately 50% higher for children among the poorest 20% of the population or households than among the least poor 20% in a group of sites constituting over half of the total. This was the finding from Matlab, Bangladesh; Agincourt, South Africa; and Ifakara and Rufiji, Tanzania – four of the five sites whose findings were selected for presentation in Table 12.1 because of their approximate comparability. (All five provided data from roughly the same period during the 1990's for under-5 (and in most cases infant) mortality, for each economically-defined quintile of the site population.) A similar finding for mortality among infants and children up to age fifteen came from Filabavi, Vietnam. There, too, mortality among the lowest 20% of the population was also about 50% above that of the top 20%.

However, there were three sites, all in West Africa, where differences in infant and child mortality across economic groups appeared to be minimal. One was Navrongo, Ghana where, as shown in Table 12.1, where deaths among infants and children under 5 occurred only 10-15% more frequently among the poorest than among the least poor 20% of the population. The Bandim, Guinea-Bissau and Farafenni, Gambia sites found no significant differences between the economic status of parents of children who died and those who did not in their case control studies – although, as the Bandim study authors note, the lack of a difference could have resulted from the case matching procedure employed.

---

dimension, this paper thus provides only a partial view of equity, and of the information about it provided through the INDEPTH studies. To the extent that such a partial approach can be justified, the justification is to be found in: the more limited prior treatment of the economic dimension in prior quantitative assessments, as explained later in the text; the somewhat greater attention given to economic dimension than to other dimensions in most INDEPTH studies; and the particularly significant contribution of the INDEPTH work to the improved understanding of equity's economic dimension.

2   Two sites participating in the project (Bandim, Farafenni) used a case control methodology that does not permit comparison with others; one (Tanzania/AMMP) limited the scope of its work to the development of a measurement method and did not provide substantive findings. Of the seven remaining sites, all of which dealt with comparisons across economically-defined groups of people, one (Watch) covers a set of small areas selected to represent the national population, so that findings reported by it (available for health service use only) cannot be taken as representing any one small geographical area. This leaves six of the ten chapters in this volume reporting on economic disparities representative of small areas.

Table 12.1   Site-level economic disparities in infant and under-5 mortality

Site and Year	Rate Ratio: Rate in Poorest Economic Quintile Divided by Rate in Least poor Quintile	
	Infant Mortality Rate	Under-5 Mort. Rate.
Bangladesh, Matlab – 1993/95 ICDDR, B Service Area Government Service Area	1.54 1.46	1.61 1.61
Ghana, Navrongo – 1996/2000	–	1.13
South Africa, Agincourt – 1992/2000	–	1.65
Tanzania, Ifakara – 2000	1.47	1.47
Tanzania, Rufiji – 1999/2000	1.46	1.53

Comparison with national inequalities   In some cases, the magnitude of the site-level infant and child mortality differences between the poorest and least poor shown in Table 12.1 are smaller than the comparable differences in the entirety of the countries where the sites are located, as would be expected given the apparently greater degree of homogeneity within the relatively small sites. But this is by no means always the case. For three of the five sites covered in Table 12.1, the intra-site differences appear approximately as large as or larger than the comparable national inequalities.

The relevant data are presented in Table 12.2. In that table, the site rate ratios shown in table 1 are reproduced on the left-hand side. On the right side of the table are country-level rate ratios produced through an analysis of data, using a similar method, from a nationwide household survey undertaken at approximately the same time that the site data were collected. (The nationwide surveys in questions were executed by the well-known Demographic and Health Survey (DHS) Program; the analyses were undertaken by the DHS secretariat in connection with a World Bank project that produced a series of country reports on health inequalities. (Gwatkin et al., forthcoming))

Because of subtle but potentially important variations in analytical approach used in producing the site and national figures, any comparison between the site and national ratios can be considered only approximate.[3] But there can be little basis

---

3   For example, many of the site studies express their findings in terms of quintiles of *households* (exceptions Ifakara and Rufiji), while the national studies present figures for quintiles of *household members* or individuals in the study population. Because poorer households tend to contain more people than better-off ones as a result of higher fertility among the poor, the poorest quintile of households usually contains significantly more than 20% of individuals in the study population; the least poor quintile of households normally contains substantially less than 20% of individuals. The result of correcting for this difference will depend upon the distribution of inequalities across different segments of the population sample, and cannot easily be identified *a priori*.

Table 12.2   Site- and national-level economic disparities in infant and under-5 mortality

Country, Site, and Year of Study	Infant Mortality Rate		Under-5 Mortality Rate	
	Site Ratio	Nat. Ratio	Site Ratio	Nat. Ratio

Bangladesh
  Matlab, 1993/95
  National, 1996/97
	1.50	1.70	1.61	1.86

Ghana
  Navrongo, 1996/2000
  National, 1998
	–	–	1.13	2.66

South Africa
  Agincourt, 1992/2000
  National, 1998
	–	–	1.65	3.99

Tanzania
  Ifakara, 2000
  National, 1999
	1.47	1.25	1.47	1.18

Tanzania
  Rufiji, 1999/2000
  National, 1999
	1.46	1.25	1.53	1.18

Notes:
– All rate ratios refer to the rate in the poorest economic quintile divided by the rate in the least poor quintile.
– For ease of presentation, the site figures presented for Matlab, Bangladesh consist of the unweighted averages for the two areas within the study site. The area- specific infant mortality ratios, as presented in summary table 1, are 1.54 for the ICDDR,B service area; and 1.46 for the government service area. The under-5 mortality ratio is 1.61 in each of the two areas.

for believing that these variations produce enough of a distortion to change the impression of a very wide range of patterns that emerges from even a quick look at the figures:

• In Agincourt, South Africa and Navrongo, Ghana, the disparities are much smaller than those found in South Africa and Ghana as a whole. The case of Ghana is particularly striking. In the country as a whole, under-five mortality in the poorest quintile is over two and a half times the magnitude of that in the poorest quintile. Within the Navrongo site, the difference appears minimal.
• In Matlab, Bangladesh, the disparities seen are quite close in magnitude to those for Bangladesh overall. For infant mortality, the Matlab rate ratio is on the order of 1.5, compared to around 1.7 for the entire country. The corresponding rate ratios for under-5 mortality are around 1.6 and 1.9, respectively. Because of the variations in methodology described in fn. 4 and the pattern of inequalities

shown in the Matlab study, the actual differences between the site and national figures are probably even smaller than suggested by the numbers, and would probably be quite small and possibly non-existent.

- In Ifakara and Rufiji, Tanzania, the intra-site differences appear distinctly larger than those for Tanzania as a whole: for both infant and under-5 mortality, the intra-site rate ratio is on the order of 1.5, compared with the national rate ratio of around 1.2. This stands in contrast to the situation found in Ghana and South Africa.

Inequalities in Health Service Coverage

The information available from the INDEPTH sites concerning inequalities in health service coverage is considerably more limited and less amenable to the development of generalizations than that for health status, for several reasons. One is that fewer of the sites covered health services in a manner that permits cross-site comparisons. Only five of the ten sites whose reports appear in this volume covered health services systematically; and one of those (Watch, Bangladesh) was a network of sites designed to produce nationally representative data rather than data for a single small area. Also, different sites focused on different types of health services, limiting the cross-site comparability of such data as exists; and each site dealt with only a few services, usually preventative services, which are not necessarily representative of the wide range of services that might exist in the site. Finally, in some cases, the sites appear to have been providing as well as measuring the coverage of services, lessening still further the comparability with those sites that were only measuring and with the record of routine, national service programs.

Nonetheless, it is clear from the annex that coverage is almost always higher among the least poor 20% of a site's population than among the poorest 20%. Only in self-treatment and treatment at the lowest-level health facilities in Filabavi, Vietnam, are coverage rates higher among the poorest than the least poor and even here, the size of the difference is marginal. For all other services in Filabavi, and for all services in the other sites, coverage was at least somewhat higher among the least poor than among the poorest. In many cases – as in the use of higher-level services in Filabavi and the possession/use of bednets in Ifakara and Rufiji, Tanzania – coverage among the least poor is over twice as high among the least poor as among the poorest.

It also appears that coverage inequalities are considerably higher for bednet possession/use than for immunization – the only two specific services covered by more than a single site. In the three sites providing coverage data on these services (Navrongo, Ghana for immunization; Rufiji, Tanzania for bednets; and Ifakara, Tanzania for immunization and bednets), bednet use or possession was typically over twice as high among the least poor 20% than among the poorest 20%; compared with an advantage of roughly 5-35% that this group enjoyed with respect to immunization. (Also, overall immunization coverage was considerably higher, which is a possible explanation for the lower inequality since large inequalities become a mathematical impossibility as universal coverage is approached.) In all cases, however, intra-site economic inequalities in immunization coverage and bednet possession/use appear consistently smaller than in the countries where the sites are located.

Possible Reasons for the Small Area Inequalities

The finding that significant inequalities in health status exist in many small areas leads naturally to the questions of why they do, or of what causes them. This is of particular relevance for economic inequalities in health status, where the evidence from the INDEPTH studies is considerably more complete – or, at least notably less incomplete – than it is for economic inequalities in health service coverage.

To provide anything like a full response to this question of what might cause such inequalities in health status within small areas lies far beyond the scope of the INDEPTH studies, and thus of this summary. However, there are enough hints in the INDEPTH material and in other, related work to suggest a few possibilities.

These possibilities fall into three broad categories:

- Economic inequalities: Perhaps large economic inequalities exist within some of the INDEPTH sites, but not in others; and that the health inequalities are caused by, or at least mirror, those economic inequalities.
- Social inequalities: Perhaps other social determinants of health status are far more powerful than economic conditions or health service use, and these other determinants are correlated with economic conditions in some sites, but not in others.
- Health service inequalities: Perhaps the failure of health interventions to reach lower economic groups as well as better-off ones are magnifying the health effects of whatever economic disparities exist in some settings.

Economic Inequalities

On the first of these possibilities, there are two pieces of suggestive information. One comes from the INDEPTH studies themselves, the other from an emerging body of statistical research that seeks to measure the prevalence of poverty in small areas. Both types of information suggest that, in many places, economic inequalities are larger than previously recognized.

Information from the INDEPTH sites   The relevant information available from the INDEPTH reports appearing in this volume is limited to data provided by the Ifakara and Rufiji, Tanzania sites. They point to the existence of clusters of people who have notably more possession than other groups. For example:

- In Ifakara, there is one group, comprising 20% or more of the population, almost all of whom live in houses with tin rather than thatched roofs, and almost all of whom have a radio and bicycle. At the other extreme is another, poorer group of equal size, none whom have any of those things.
- In Rufiji, everybody in one group of extremely poor people, also representing 20% or more of the population, lives in a house with a thatched roof and earthen floor. But there's another, equally large better-off group whose members are 70-80% likely to live in some improved type of house, with a cement floor and/ or an asbestos roof.

Certainly, a roof with a tin or asbestos rather than a thatched roof is hardly a mansion; and, in the absence of multivariate analysis or other information, there is no way of knowing for sure that this difference in living standards was in itself was responsible for the health inequalities recorded. But is certainly possible that this difference, while too subtle to attract much notice from an outside observer, was quite large in local terms; and, when taken together with additional economic disparities that it no doubt denotes, was sufficient to produce large health inequalities.

Information from poverty prevalence research   The body of statistical research that provides additional relevant information is being developed by economists interested in identifying the level of poverty that exists in small areas, in order to permit more precise geographic targeting. The method employed, sometimes referred to as 'poverty mapping' is similar to that employed in the AMMP paper appearing in this volume. Its objective is to overcome the difficulty in estimating the prevalence of poverty in small geographic areas caused by the absence of expenditure or income data in most surveys, such as national censuses, large enough to provide information about small areas throughout a country.

Poverty mappers start with data from a (usually) small household expenditure survey that also contains information about a large number of factors that can potentially explain differences in levels of expenditure observed. The mappers take those of the potential explanatory factors that are also covered by a national census or similar study and work out the statistical relationship between them and expenditure. These relationships are then applied to data about the explanatory factors for each of a country's households provided by the census, in order to estimate or predict expenditures. The resulting predictions, while rarely reliable at the level of the individual households, usually can produce workably accurate community-level figures. This permits production of a poverty 'map' – or to be more accurate, list of communities indicating the percentage of households in each whose expenditures lie below a specified poverty line.[4]

Application of this technique remains in its infancy, and there is unfortunately no known poverty map for any of the countries covered by the INDEPTH studies. However, the studies appearing to date have found that around half or more of total inequalities found within such varied settings as Ecuador, Madagascar, and Mozambique are attributable to disparities within communities (Elbers et al., 2003). This is in line with earlier, less formal work suggesting less dramatic, but nonetheless significant local-level disparities in Mexico (Baker and Grosh, 1994), India, and Romania (Gwatkin, 1998).

It is thus beginning to appear possible that intra-community disparities in economic status are considerably larger and more frequent than earlier believed. The common existence of such disparities would go a long way toward explaining the INDEPTH findings; and, conversely, the INDEPTH findings constitute yet one more indication that such intra-community economic disparities frequently exist.

---

4   A fuller introduction to this approach can be found at the World Bank's poverty mapping website: http://web.worldbank.org/WBSITE/EXTERNAL/TOPICS/EXTPOVERTY/ EXTPA/0,,contentMDK:20219777~menuPK:462078~pagePK:148956~piPK:216618~ theSitePK:430367,00.html

Social Inequalities

The second possibility is that the economic disparities just described, even if real, bear limited if any responsibility for the observed inequalities in health status. Rather, the economic disparities are simply correlated with other, social disparities that are the health inequalities' true cause.

This possibility comes through most clearly in the study from the Agincourt, South Africa site. The Agincourt investigators found a clear inverse relationship between under-five mortality and economic status. But, not surprisingly, they found that under-five mortality was related to a lot of other things as well: gender and nationality of the household head; mother's marital status and educational attainment; and others. Although the study includes only limited use of the multivariate analysis often applied in an effort to assign causality to a particular factor, the statistical relationships and supplementary qualitative information presented appear to point toward nationality of household head as especially important determinant. As explained in the study report, Agincourt lies close to the border with Mozambique, whose civil war had led many Mozambicans to flee into South Africa. These refugees, who constituted nearly a third of the Agincourt population, were perhaps better off than they would have been had they stayed at home. But in South Africa they remained marginalized, living in communities largely without education and health services, and without the legal residency needed to protect them from economic exploitation. So while economic disparities in health status may exist, to focus unduly on them would cause one to miss the basic point that such disparities appear to be simply one of many manifestations of poor living standards whose basic cause is likely to be national origin.

Also noteworthy are the efforts of the Bandim, Guinea-Bissau and Farafenni, Gambia site investigators to explore other possible causes or dimensions of health inequalities. Of particular interest to them was social capital, or the strength of social support available from other community members. As it turned out, neither investigation uncovered indications of health inequalities by either economic status or degree of social support. But both nonetheless stand as valuable reminders of the multidimensional and complex nature of poverty. They also serve as warnings against assuming either that the existence of economic inequalities in health mean that the economic inequalities are an important cause of the health disparities; or that the absence of economic inequalities in health means that all is well, since important health inequalities could still exist with respect to some other, equally significant dimension of poverty.

In other words, in looking at economic inequalities in health, one must take particular care to heed the standard warning against inferring causality from associations found in statistical studies. Economic inequalities, like other standard inequalities used in health equity work (such as gender, ethnicity, etc.), are no more than bivariate associations between economic and health status, which are intended to identify and measure one dimension of an inequality thought to be inequitable, with no implication of causality.

Health Service Inequalities

The third possibility, which might help explain at least some of the health status inequalities reported in Annex 12.1, is that those inequalities are exacerbated by the inequalities in health service coverage reported in Annex 12.2 and discussed earlier. To the extent that health services exercise an influence on health status, as one would hope that they do, it would seem reasonable to expect the health service inequalities to be at least one of the factors responsible for the health status inequalities observed.

Certainly, significant inequalities in health service coverage appear nearly ubiquitous. They have been demonstrated to exist time and time again at the national level, not just for higher-level services but also for basic service programs directed particularly toward disadvantaged groups (Gwatkin et al., 2000 and forthcoming). The limited evidence from the INDEPTH sites covered in Annex 12.2, and summarized in the preceding section, suggest that large coverage inequalities also exist for many, if not necessarily all, services in small areas within countries. What remains to be determined is how large and how frequent these coverage inequalities are for how many services; and whether the therapeutic benefit of the health services in question is strong enough to influence significantly the health status of those receiving them.

Contribution Two: Helping Confirm the Feasibility of Studying Economic Inequalities in Health

Prior Neglect of Health Equity's Economic Dimension

Until recently, equity issues had attracted only spotty attention from demographic, epidemiological, and other health specialists engaged in field research. While some dimensions had been fairly well covered, others had been almost completely ignored.

By far the most attention had been given to gender. Demographic research in particular almost always differentiated between males and females in the construction of population estimates, which produced such important findings as higher life expectancy among females in most parts of the world outside the Middle East and South Asia. Often, survey instruments also covered issues like educational level, occupation and place of residence; and this made it possible to note that better educated people in higher-level occupations and living in cities usually had better health those who are less well educated, work in menial jobs, and in the countryside.

However, the economic dimension of health inequality had rarely been covered. Since, for better or worse, inequalities in economic status arguably occupy an especially prominent significance in the popular consciousness, the absence of information about it was particularly noticeable.

As noted in the chapter by Saul Morris appearing elsewhere in this volume, an important reason for this absence had been the difficulty in gathering information about income or consumption, the two most commonly used measures of economic status. The amount of time and effort to collect accurate data on either of these measures has made it impractical to consider including them in health surveys, with instruments already overburdened with the numerous and complicated questions needed to collect accurate information about the topics of primary interest.

Use of Wealth as an Indicator of Economic Status

Thanks to the recent developments described by Morris, the situation has changed. Of particular importance has been recognition that wealth can be used in place of income or consumption an indicator of economic status. Usually, wealth correlates closely enough with income or consumption to justify using it as a proxy for them. Even when it does not, wealth can justifiably be used as an indicator that captures a dimension of economic status that differs from, but is equally important as income or consumption.

While wealth is not without important conceptual and measurement issues, it has the critical advantage of being far easier to determine than either income or consumption (see Annex 12.3). Usually, some 8-10 questions – about such household attributes as type of roofing and/or flooring material, source of water, presence of electricity, possession of a watch and/or radio – suffice to produce an index of material capital that, while conceptually limited, can be shown to track rather well with income or consumption. Such questions normally require no more than 2-3 minutes to ask, entail no recall, often involve items that are readily visible and thus easily verifiable, and deal with topics that are rarely considered sensitive by the person interviewed. Since many of the questions are often included already in health or demographic household surveys, the amount of extra time required ask enough additional ones to construct a viable asset index is normally minimal.

Contribution of the INDEPTH Studies

At the time when work began on the papers appearing in this volume, the wealth or asset approach had been applied with apparent success by a few highly skilled and experienced researchers working primarily with large-scale data sets collected by leading Northern institutions.[5] But the methods used still remained largely untried, and it was by no means clear how widely they could be applied. In particular, the question remained open whether the household asset approach could be successfully applied to the analysis of data from smaller surveys with fewer household asset questions, or by investigators without extensive prior experience with the approach.

The INDEPTH experience points toward an affirmative response to the question just posed. The information needed for the application of asset index approach turned out either to exist already in the data bases of the INDEPTH sites, or to be amenable to collection through a manageable degree of extra effort during one of the site's regular survey rounds. Since none of the investigators had previously worked with an asset index, there was need to provide an introduction through a series of technical workshops. But once this was done, the data analysis required proved to be well within the range of statistical competence existing in most of the sites. The sites'

---

5   Examples include the pioneering assessment of education differentials by Deon Filmer and
    Lant Pritchett, using data from the U.S. AID-supported Demographic and Health Survey
    (DHS) program Filmer and Pritchett, 2000); the several studies by Adam Wagstaff based
    on the World Bank's Living Standards Measurement Surveys (e.g. Wagstaff, 2000); and
    assessments of DHS data for Africa by David Stifel, David Sahn, and Stephen Younger
    (Stifel, Sahn, and Younger, 1999).

sample sizes usually proved adequately large to detect differences among economic groups in small, seemingly homogeneous geographic areas.

All this suggests that the asset or wealth approach has the potential to become a standard instrument for use in measuring economic inequalities in health service use and status through epidemiological field research and intervention monitoring programs. To be sure, it still has a number of rough edges that need smoothing – through, for example, broader agreement about just which asset questions should be included in a survey instrument, about how the different assets should be weighted in the construction of an index, or about which inequality indicators should be used in the presentation of results. Further, the technique's widespread application is likely to require considerably more workshops or other opportunities than currently exist for investigators to become familiar with it. But none of these challenges seem insurmountable.

Thus, the experience gained through this exercise points to the feasibility of aspiring to assess regularly economic as well as other inter-group inequalities in health status and service use in the great majority of, if not all, sites belonging to the INDEPTH network and other sentinel surveillance programs covering populations of, say, 50,000 or more. Sites and surveys covering smaller populations should be equally able to measure economic and other inequalities in relatively common events like the prevalence of malnutrition or coverage of basic health programs. And if this can be done, it would seem equally feasible to consider regularly monitoring not just the degree of overall coverage achieved by large-scale health interventions, but also the economic profile of those covered in order to determine how well their interventions are reaching the poor.

Conclusion

Given the focus of the INDEPTH project on health equity, it would be appropriate to close by considering briefly the implications of the findings and experiences reported in this volume for efforts to improve conditions among particularly disadvantaged population groups. This can best be done through a quick reprise, from a policy perspective, of the INDEPTH project's principal contributions presented at the outset and developed in the text.

Limitations in the Potential Effectiveness of Geographic Targeting

The first contribution – the demonstration of frequent inequalities even in small areas – points toward possibly important limitations in the potential effectiveness of geographic targeting techniques in reaching the neediest people. Targeting any specific geographic area for the provision of services can work in reaching disadvantaged groups only to the extent that inequalities exist between rather than within the areas being considered as possible targets. The INDEPTH finding of significant inequalities within small areas thus throws into question the value of selecting those areas as a basis for program emphasis, since one might end up simply reaching the better-off in those areas rather than the truly needy. This is not necessarily an issue everywhere: in some places, such as Ghana, small area differences seem modest in relation to the very large differences existing between areas, suggesting that geographic targeting might perhaps work reasonably well. But elsewhere, as in Southern Tanzania and possibly rural Bangladesh, large intra-site disparities raise doubts about geographic targeting's likely power. This calls for looking considerably more carefully than in the past at the spatial distribution of poverty before adopting geographic targeting as a central element in a strategy for reaching high-priority households.

Strengthened Case for Routinely Assessing Health Programs from an Equity Perspective

The second contribution – helping confirm the feasibility of studying economic inequalities in health – significantly strengthens the case for routinely assessing health programs from an equity as well as an effectiveness perspective. The INDEPTH studies demonstrate that such routine assessments are considerably less challenging than previously thought, because of the feasibility of collecting and analyzing asset/ wealth information under field conditions. This sets the stage for assessments that appear likely to prove especially valuable in gauging how well equity-oriented health service initiatives achieve their stated but often unachieved goal of reaching not simply large numbers of people, but large numbers of the people who need them most.

References

Baker, July L. and Margaret E. Grosh. 'Measuring the Effects of Geographic Targeting on Poverty Reduction', Living Standards Measurement Study Working Paper No. 99. Washington, D.C: The World Bank, (1994).

Elbers, Chris, Peter Lanjouw, Johan Mistiaen, Berk Özler and Ken Simler. 'Are Neighbours Equal? Estimating Local Inequality in Three Developing Countries?', Discussion Paper No. 2003/52, World Institute for Development Economics Research, United Nations University, (2003). (available at: http://econ.worldbank.org/files/21898_enotedone2.pdf)

Filmer, Deon and Lant Pritchett. 'Estimating Wealth Effects without Expenditure Data-or Tears: An Application to Educational Enrollments in States of India'. *Demography*, vol 38, no. 1 (February 2001), pp. 115-132.

Gwatkin, Davidson R. 'Poverty, Equity, and Health in the Developing World: An Overview', Washington, D.C.: The World Bank, (1998).

Gwatkin, Davidson R., Shea Rutstein, Kiersten Johnson, Eldaw Abdalla Suliman and Adam Wagstaff. Socio-Economic Differences in Health, Nutrition, and Population, 2nd Ed. Washington, D.C.: The World Bank, forthcoming. (available at: http://www1.worldbank.org/prem/poverty/health/data/round2.htm)

Stifel, David, David Sahn and Stephen Younger. 'Inter-temporal Changes in Welfare: Preliminary Results for Nine African Countries', CFNPP Working Paper no. 94, Cornell University, 1999 (available at: http://www.he.cornell.edu/cfnpp/images/wp94.pdf)

Wagstaff, Adam. 'Socioeconomic Inequalities in Child Mortality: Comparisons across Nine Developing Countries'. Bulletin of the World Health Organization, vol 78, no. 1 (January 2000), pp. 19-29.

Annex 12.1   Economic inequalities in health – health status

Health Status Indicator	Low-High Rate Ratio		
	Rate in Poorest 20%	Rate in Least Poor 20%	Rate Ratio
Matlab, Bangladesh (ICDDR,B Service Area)			
Under-5 Mort.Rate, All Causes (1983/85)	157.3	118.1	**1.33**
Under-5 Mort.Rate, All Causes (1993/95)	87.7	54.6	**1.61**
Infant Mort.Rate, All Causes (1983/85)	103.8	93.0	**1.12**
Infant Mort.Rate, All Causes (1993/95)	67.1	43.7	**1.54**
1-4 Mort.Rate, All Causes (1983/85)	53.5	25.1	**2.13**
1-4 Mort.Rate, All Causes (1993/95)	20.6	10.9	**1.89**
Under-5 Mort.Rate, Diarrhoea (1983/85)	33.3	24.0	**1.39**
Under-5 Mort.Rate, Diarrhoea (1993/95)	14.1	6.2	**2.27**
Under-5 Mort.Rate, Pneumonia (1983/85)	15.4	14.5	**1.06**
Under-5 Mort.Rate, Pneumonia (1993/95)	8.9	11.0	**0.81**
Under-5 Mort.Rate, Other Causes (1983/85)	105.6	78.6	**1.34**
Under-5 Mort.Rate, Other Causes (1993/95)	64.0	36.5	**1.75**
Matlab, Bangladesh (Government Service Area)			
Under-5 Mort.Rate, All Causes (1983/85)	211.8	151.9	**1.39**
Under-5 Mort.Rate, All Causes (1993/95)	136.8	85.2	**1.61**
Infant Mort.Rate, All Causes (1983/85)	131.4	108.5	**1.21**
Infant Mort.Rate, All Causes (1993/95)	97.7	66.8	**1.46**
1-4 Mort.Rate, All Causes (1983/85)	80.4	43.4	**1.85**
1-4 Mort.Rate, All Causes (1993/95)	39.1	18.4	**2.13**

Annex 12.1 (Continued)

Health Status Indicator	Low-High Rate Ratio		
	Rate in Poorest 20%	Rate in Least Poor 20%	Rate Ratio
Under-5 Mort.Rate, Diarrhoea (1983/85)	49.5	28.4	**1.74**
Under-5 Mort.Rate, Diarrhoea (1993/95)	22.4	13.2	**1.70**
Under-5 Mort.Rate, Pneumonia (1983/85)	16.5	13.5	**1.22**
Under-5 Mort.Rate, Pneumonia (1993/95)	26.1	9.7	**2.69**
Under-5 Mort.Rate, Other Causes (1983/85)	138.9	107.2	**1.30**
Under-5 Mort.Rate, Other Causes (1993/95)	86.8	56.2	**1.54**
Navrongo, Ghana			
Percent of Under-5 Children Dying	14.2	12.6	**1.13**
Agincourt, South Africa			
Under-5 Mortality Rate	27.5	16.5	**1.67**
Ifakara, Tanzania			
Under-5 Mortality Rate	188	138	**1.36**
Infant Mortality Rate	109	74	**1.47**
Child Mortality Rate	89	69	**1.29**
Rufiji, Tanzania			
Under-5 Mortality (deaths/1,000 under-5 person-years observed)	36.7	24.0	**1.53**
Infant Mortality Rate (deaths/1,000 infant person-years observed)	107.1	73.6	**1.46**
Child Mortality Rate (deaths/1,000 child person-years observed)	14.3	7.7	**1.86**
FilaBavi, Vietnam			
Age-Standardized Mort. Rate, under 15 Years	2.88	1.89	**1.52**
Age-Standardized Mort. Rate, 15-59 Years	4.89	1.38	**3.54**
Age-Standardized Mort. Rate, over 60 Years	35.50	32.01	**1.02**

Annex 12.2   Economic inequalities in health – health service use

Health Service Use Indicator	High-Low Rate Ratio		
	Rate in Poorest 20%	Rate in Least Poor 20%	Rate Ratio
Watch, Bangladesh[6]			
Percent of Women Receiving Antenatal Care	15.7	52.2	**3.32**
Percent of Women Receiving Tet.Toxoid Imm.	59.0	81.6	**1.38**
Percent of Women Receiving Hosp.Delivery	1.2	9.0	**7.50**
Percent of Women Receiving Postnatal Care	7.2	25.9	**3.60**
Percent of Women Receiving at Least 1 Ser.	62.2	85.9	**1.38**
Percent of Women Receiving at Least 2. Sers.	16.8	55.3	**3.29**
Percent of Women Receiving at Least 3 Sers.	3.9	23.5	**6.03**
Percent of Women Receiving 4 Services	0.4	4.7	**11.75**
Navrongo, Ghana			
Percent of Children Immun. against Measles	73.5	81.9	**1.11**
Percent of Children Fully Immunized	48.8	65.6	**1.34**
Ifakara, Tanzania			
Percent of Children Living < 1 Hr. from H.Fac,	50	58	**1.16**
Percent of Children Receiving BCG Vaccine	83	89	**1.07**
Percent of Children Sleeping under Bednet	29	76	**2.62**
Percent of Households with 1+ Bednets, 1997	23	60	**2.61**
Percent of Households with 1+ Bednets, 2000	53	93	**1.75**

---

6   Watch data come from a set of sites in different parts of Bangladesh that are intended to provide a representative view of the country as a whole. This limits the comparability between the Watch data and the data from the other sites presented here, which come from single, small homogeneous areas that do not seek to reflect the wider range of conditions found in the countries where they are located.

Annex 12.2 (Continued)

Health Service Use Indicator	High-Low Rate Ratio		
	Rate in Poorest 20%	Rate in Least Poor 20%	Rate Ratio
Rufiji, Tanzania			
Percent of Households with 1+ Bednets	5.7	21.2	**3.72**
FilaBavi, Vietnam			
(All figures are unweighted averages for male and female)			
Use of Commune Health Station	4.86	4.45	**0.92**
Use of District Health Center	2.92	4.38	**1.50**
Use of Provincial Hospital	0.90	2.30	**2.56**
Use of Traditional Healer	1.84	2.12	**1.15**
Use of Self-Treatment	50.78	49.60	**0.98**
Use of Private Facilities	30.00	32.42	**1.08**

Annex 12.3 Types of asset included in household wealth assessments in various studies

Asset	Rufiji, Tanzania	Agincourt, South Africa	Filabavi, Vietnam	Matlab, Bangladesh	Navrongo, Ghana	Watch, Bangladesh	Farafenni, Gambia	Ifakara, Tanzania	Bandim, Guinea Bissau	Frequency
Property										
Radio or audio tape recorder	●	●	●	●	●	●	●	●		8
Television	●	●	●	●		●			●	6
House ownership	●			●			●	●		4
Bed	●			●	●		●			4
Wardrobe	●		●	●		●				4
Clock/watch	●			●		●				3
Sewing machine	●		●		●					3
Refrigerator	●	●	●							3
Video Cassette Recorder	●	●	●							3
Land ownership	●		●							2
Satellite dish	●	●								2
Lamp				●	●					2
Mattress/Blanket	●			●						2
Table				●		●				2
Fan	●		●							2
Mobile phone		●								1
Telephone		●								1
Iron	●									1
Sofa	●									1
Stove		●								1
Livestock										
Cow, Sheep, Goat, or Pig	●	●	●	●	●	●	●	●	●	9
Poultry	●	●				●		●		4
Transportation										
Bicycle	●	●	●			●	●	●		6
Motorcycle or motorboat	●	●	●	●						4
Automobile	●	●	●							3
Donkey carts		●			●					2
Energy										
Type of energy source	●	●		●	●	●			●	6
Water & Sanitation										
Type of water supply	●	●	●	●	●	●				6
Type of sanitation	●	●	●	●					●	5
Housing Characteristics										
Type of roofing material	●	●	●	●	●		●	●	●	8
Type of wall material	●	●	●	●						4
Number of rooms	●	●			●				●	4
Type of flooring material	●	●	●							3
Frequency	27	21	17	16	11	9	6	6	6	

Note: Type of energy, water supply, sanitation, roofing, flooring, and walls typically included about 5 locally relevant types for each.